NEUROSCIENCE RESEARCH PROGRESS

CHRONIC DISORDERS OF CONSCIOUSNESS

FROM RESEARCH TO CLINICAL PRACTICE

NEUROSCIENCE RESEARCH PROGRESS

Additional books in this series can be found on Nova's website under the Series tab.

Additional e-books in this series can be found on Nova's website under the e-book tab.

NEUROSCIENCE RESEARCH PROGRESS

CHRONIC DISORDERS OF CONSCIOUSNESS

FROM RESEARCH TO CLINICAL PRACTICE

ROCCO SALVATORE CALABRÒ
AND
PLACIDO BRAMANTI
EDITORS

New York

For permission to use material from this book please contact us:
Telephone 631-231-7269; Fax 631-231-8175
Web Site: http://www.novapublishers.com

Library of Congress Cataloging-in-Publication Data

Library of Congress Control Number: 2013939115

ISBN: 978-1-62618-940-9

Published by Nova Science Publishers, Inc. † New York

To Mario and Giorgia
my children,
my whole life!

"Children have neither a past nor a future. Thus they enjoy the present, which seldom happens to us"
Jean de La Bruyere

"Learning is acquired by reading books; but the much more necessary learning, the knowledge of the world, is only to be acquired by reading men and studying all the various editions of them"
P. Chesterfield

"It is impossible to live pleasurably without living prudently, honorably, and justly; or to live prudently, honorably, and justly, without living pleasurably"
Epicuro

Contents

Preface

The 20th century witnessed tremendous scientific progress in many fields. This has brought about a better understanding of the world we inhabit, of the forces that drive it, of the relationships between the human race and the rest of the universe. Scientific explanations have been provided for most of the phenomena that used to be considered divine events. We have learned how the universe was born, and how it gave rise to the galaxies and the stars and ultimately to our planet; and what life is, how it survives, reproduces and evolves; and what the structure of the brain is, and how it works. The mystery is no longer in our surroundings: it is inside ourselves!

What we still cannot explain is precisely that: "ourselves". We may have a clue to what generates reasoning, memory and learning. But we have no scientific theory for the one thing that we really know very well: our consciousness, our awareness of being us, ourselves.

Perhaps no aspect of mind is more familiar or more puzzling than consciousness and our conscious experience of self and world. The problem of consciousness is arguably the central issue in current theorizing about the mind. Despite the lack of any agreed upon theory of consciousness, there is a widespread consensus that an adequate account of mind requires a clear understanding of it and its place in nature.

An animal, person or other cognitive system may be regarded as conscious in a number of different senses:

- *Sentience:* it may be conscious in the generic sense of simply being a *sentient* creature, one capable of sensing and responding to its world; being conscious in this sense may admit of degrees, and just what sort of sensory capacities are sufficient may not be sharply defined.

- *Wakefulness: this* further require that an organism might be considered as conscious only if it is *awake and normally alert;* in that sense organisms would not count as conscious when asleep or in any of the deeper levels of coma; again boundaries may be blurry, and intermediate cases may be involved.
- *Self-consciousness:* a third and yet more demanding sense might define conscious creatures as those that are not only aware but also aware that they are aware, thus treating creature consciousness as a form of *self-consciousness*; the self-awareness requirement might get interpreted in a variety of ways, and which creatures would qualify as conscious in the relevant sense will vary accordingly. If it is taken to involve explicit conceptual self-awareness, many non-human animals and even young children might fail to qualify, but if only more rudimentary implicit forms of self-awareness are required then a wide range of nonlinguistic creatures might count as self-conscious.

Thus, consciousness is to be considered an ambiguous term, encompassing both wakefulness and awareness. This distinction is crucial to the concept of the vegetative state (VS), in which wakefulness recovers after brain injury without recovery of awareness.

Chronic unconsciousness (DOC) is a tragic and ironic failure of high-technology treatment to preserve or restore brain function, the primary aim of therapeutics. Management of a patient in a VS or a minimally conscious state (MCS) requires carefully reaching the correct diagnosis, pronouncing an evidence-based prognosis, and thoughtfully considering the medical, ethical, and legal elements of optimum treatment.

In the USA, the case of Theresa Schiavo in 2005 brought many of these issues to intense public scrutiny, albeit in an unfortunately sensationalised and politicised way.

Patients might be in a VS or MCS temporarily as a stage during recovery from a traumatic brain injury or a non-traumatic brain insult, or might reside in the state chronically and permanently. The issue of prognosis for recovery of awareness is a crucial but difficult clinical determination.

As it has repeatedly emphasized, ultimately we can only witness behavior, and we have no access to other people's first-person subjective experiences. Furthermore, even though a conscious experience may give rise to a certain behavior, we cannot deduce from our observations of acts of behavior the presence of a mental life.

How then are we to know that our fellow human beings are conscious? At the very best we can infer from other people's behaviors that they are conscious. If throughout our lifetimes we consistently see that people exhibit the same responses to environmental events as ourselves, we must conclude that their experiences must also be very similar to ours. So in a sense, we are continuously trying to read other people's minds, based on their behavior and verbal output.

In essence, they take as their starting point the patient's neural behavior, and use these observations to draw inferences about the contents of their mental life. This approach is motivated by the spectacular amount of knowledge accumulated in the field of cognitive neuroscience. This field of research can be considered an attempt at identifying how neural states map onto mental states and/or behavior states.

The current state of the field of cognitive neuroscience allows us nowadays to tell with a reasonable amount of certainty what is going on in someone's mind, based solely on inspection of his or her neural activity.

Thus, assessing self-consciousness in coma survivors who remain unable to verbally or non-verbally express their thoughts and feelings is difficult by means of behavioral observation only. Self-referential stimuli such as the patient's own name and own face are clinically valuable arousing and attention-grabbing emotional stimuli and their use increase the chances to obtain non-reflex (''willed'' or ''voluntary'') motor responsiveness in DOC. Event-related potential and functional neuroimaging studies using such stimuli are currently being validated to help in differentiating unconscious vegetative from fluctuating minimally conscious patients. At present, based on evidence obtained from neuropsychological, neuropathological, electrophysiological and neuroimaging studies in sleep, pharmacological coma, pathological coma and related disorders of consciousness, we must conclude that the neural underpinnings of self-referential ''unconscious'' vs ''conscious'' processing remain to be further elucidated prior to their use in non-communicative brain damaged patients.

This book is aimed at examining the complex pathophysiology of consciousness disorders with their multifaceted assessment and treatment, including new diagnostic tools and novel prognostic markers, and the important bioethical issues concerning "the end of life" and the "dignity of the person".

Rocco Salvatore Calabrò
MD, PhD

Corresponding Authors

Chapter 1. Anatomo-physiology of consciousness
salbro77@tiscali.it

Chapter 2. The wide spectrum of consciousness disorders
salbro77@tiscali.it

Chapter 3. Clinical and instrumental evaluation of chronic disorders of consciousness
seregen@tiscali.it

Chapter 4. Neuroimaging assessment in low levels consciousness patients
silvimarino@gmail.com

Chapter 5. New diagnostic tools and markers of recovery in consciousness disorders
silvimarino@gmail.com

Chapter 6. The pharmacotherapy of "awakening"
salbro77@tiscali.it

Chapter 7. Rehabilitation and nursing, care-givers and psychological support
silvimarino@gmail.com

Chapter 8. Nutrition and other medical problems in VS and MCS
danielafloridia@yahoo.it

Chapter 9. Rehabilitation Paths in Severe Brain Injury
fabrizio.denaro@gmail.com

Chapter 10. Legal and Bioethical issues in chronic consciousness disorders
salbro77@tiscali.it

In: Chronic Disorders of Consciousness ISBN: 978-1-62618-940-9
Editors: R. S. Calabrò and P. Bramanti

Chapter 1

Anatomo-Physiology of Consciousness

Rocco Salvatore Calabrò and Placido Bramanti
IRCCS Centro Neurolesi "Bonino-Pulejo", Messina, Italy

Abstract

A basic understanding of consciousness and its neural correlates is of major importance for all clinicians, especially those involved with patients with altered states of consciousness. In this chapter it is shown that consciousness is dependent on the brainstem and thalamus for arousal; that basic cognition is supported by recurrent electrical activity between the cortex and the thalamus at gamma band frequencies; and that some kind of working memory must, at least fleetingly, be present for awareness to occur.

Introduction

Consciousness is a multifaceted concept that has two major components: awareness of environment and of self (i.e. the content of consciousness) and wakefulness (i.e. the level of consciousness).

Whereas the level of arousal reflects the overall state of activity in the brain, conscious awareness is a more dynamic and complex process involving various cerebral networks at any one time. There are several competing theories on how we become aware of environmental and internal stimuli, although it is widely believed to depend on interactions between the cortex and specific and non-specific (eg, intralaminar) thalamic nuclei [1,2].

Conscious awareness and arousal states also interact. Without arousal, there is no awareness, and in states of high arousal, awareness can be focused on one modality at the expense of others. To this end, animal models have been used to demonstrate that in addition to a core generalized arousal, there are also more specific forms of arousal, such as hunger, sexual behavior, and fear, which enhance responses to specific stimuli. Conversely, awareness also influences arousal, such as the abrupt increase in arousal when an alarm goes off [3].

Generally speaking, there are two opposing views about the extensiveness of neuronal involvement in consciousness. The *holistic approach* to consciousness highlights the perspective that all neurons in the brain collectively form the neural correlate of consciousness that consequently cannot be localized to a specific set of neural cells. On the other hand, the *neuronal specificity* approach supports the idea that consciousness depends on the formation of complex arrangements which can be related to specific groups of neural cells; thus, consciousness is believed to be a product of the processing of multiple brain regions, which in turn produce a high level of information integration [4].

Neuronal Circuitries Subserving Consciousness

The initial discovery that specific brain areas could drive overall cerebral activity appeared in the work of Moruzzi and Magoun [5]. These investigators proposed the existence of an ascending reticular activating system (ARAS) in the upper brainstem tegmentum (reticular formation) and central thalamus that promoted widespread cortical activation [5-6].

It has subsequently showed that the ARAS is not a monolithic activating system, but a group of interdependent subcortical and brainstem areas having specific roles in arousal and awareness. Interestingly, the core areas for maintaining an awake state seem to be the cholinergic neurons in the dorsal tegmentum of the midbrain and pons. Indeed, these areas activate the central thalamus (primarily intralaminar nuclei) and basal forebrain, which consequently stimulate the cortex through glutamatergic and cholinergic projections, respectively. In addition to supporting arousal, the basal forebrain is active during REM sleep and the central thalamus plays a role in conscious awareness.

Other brain regions, including the nuclei in the upper brainstem acting on the basal forebrain, thalamus, hypothalamus, striatum, and associative cortex are also involved in arousal but have a more modulatory role [7].

The consciousness system also comprises cortical regions important for higher-order integration including the lateral frontal and parietal association cortex, as well as the medial frontal, anterior, and posterior cingulate, and medial parietal (precuneus, retrosplenial) cortex. Indeed, stimuli are not acted on as reflexes, but typically require motivation to enter conscious awareness through a process called intention or goal-directed behavior. It has been suggested that dopaminergic modulation of glutamatergic inputs from the cortex and thalamus onto medium spiny neurons in the striatum contributes to cognition and the expression of consciousness [8].

Amusingly, lesion and functional brain imaging studies demonstrate that goal-directed behavior is primarily driven by the medial frontal and anterior cingulated cortices. These cortical regions are supported in producing goal-directed behavior by striato-pallidal-thalamic loops as well as the ventral tegmental area and the periaqueductal gray of the brainstem. The arousal systems discussed previously also act directly on the goal-directed behavior network, primarily through innervation of the striatum and frontal cortex.

Overall, the large number of regions involved in arousal provide redundancy, so that selective damage to one region, even if bilateral, only rarely results in permanent unconsciousness. However, it is important to underline that, although the content of consciousness is associated with activity of the cerebral cortex, the reticular activating system is critical for the existence of the conscious state.

Neuro-Physiology of Consciousness

Consciousness entails desynchronized, low-amplitude brain activity ranging from 20 to 70 Hz. Embedded in this pattern of global desynchronization are phase-locked, synchronized areas, which progressively oscillate in unison at the gamma band frequency of 40 Hz, and are believed to be coordinated by the intralaminar nuclei of the thalamus. Nevertheless, the information flowing back and forth between the intralaminar thalamic nuclei and the cortex through the thalamo-cortical system is believed to set up the 40 Hz oscillating loop. The neurons of the definite thalamic nuclei have been shown to activate specific cortical pyramidal cells and silence other brain areas via activation of 40 Hz gamma-aminobutyric acid (GABA) inhibitory interneurons. Thus, consciousness is supposed to be possible only when the 40 Hz electrical hum is sustained among the brain circuits, with the recruitment of more cerebral networks allowing our consciousness to widen. Significant coupling in the gamma frequency has been demonstrated to take place in both wakefulness and rapid-eye movement (REM) sleep, although the variability of gamma cycle duration is larger during sleep. Sensory perturbation can reset this gamma rhythm in awake but not sleep states, signifying that the cortex activates itself from within during sleep. During deep sleep, however, the intralaminar thalamic nuclei are inactive and no 40 Hz oscillations arise [9].

Notably, even at the lowest level of consciousness, the different areas considered essential for consciousness cannot function in isolation. Thus, it is logical to believe that consciousness involves widespread brain activity, although this does not necessarily imply homogenous activity throughout the brain. Indeed, during consciousness, brain activity is distributed throughout diffuse regions of the brain, while in unconscious states activity is supposed to remain localized [10]. Although it is generally accepted that the different features of objects are processed in different areas of the brain, and that the different qualities are cognitively bound together and experienced against the background of previous experiences, the matter of how this binding is made remains. At present, there are three hypothesized solutions to the binding problem, i.e. binding by convergence, binding by assembly and binding by synchrony, but it is possible that the final binding product is due to an interaction between all three modes of binding.

The first, i.e. *binding by convergence*, states that information from lower-order neurons is bound by higher order ones; this set of higher-order neurons then fires as an integrated binding unit once the full set of inputs have been received. The second, i.e. *binding by assembly*, supports the idea that a binding unit consists of a Hebbian cell assembly that interconnects a distinct set of neurons, through increases in synaptic efficiency within a circuit of neurons, occurring due to repeated and persistent presynaptic stimulation of the postsynaptic cells. The third possibility, i.e. *binding by synchrony*, supports the idea that the correlation between neurons occurs in the temporal dimension. As abovementioned, it is believed that distinct neural subpopulations of the cortex are synchronized due to 40 Hz oscillations of the thalamus. Binding therefore occurs due to temporal coherence of thalamo-cortical activation of pyramidal cells, with the activation returning to the thalamus in a resonant thalamo-cortical reactivation loop. This approach gives rise to more flexibility than both binding by convergence and assembly and is believed to entail a virtually endless coding capacity [11, 12].

Brain and Consciousness

Herein are more specifically described the main structures involved in maintaining alertness (figure 1), and consciousness (figure 2).

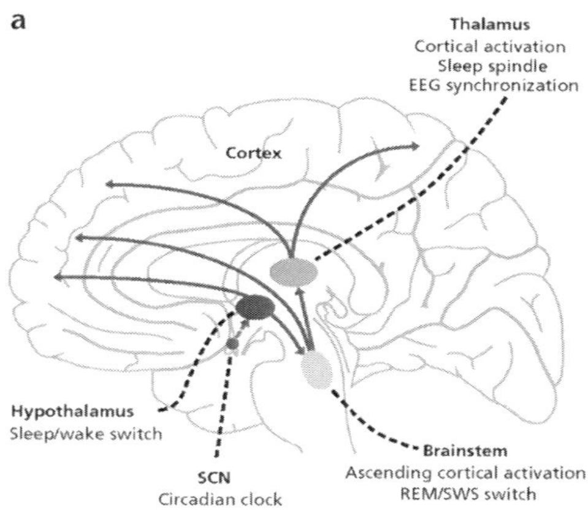

Figure 1. Shows the Reticular ascendent system and its necessary projections in activating awareness.

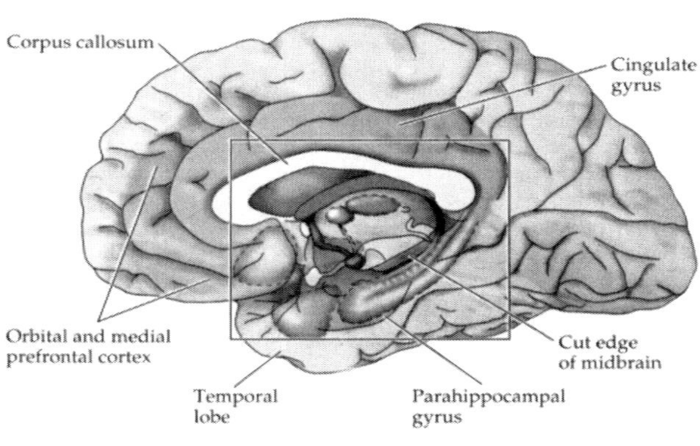

Figure 2. Shows the main brain areas involved in consciousness.

1. The Reticular Activating System

Activity of the cerebral cortex is dependent upon both specific sensory input and non-specific activating impulses from the brain stem. The source of these activating impulses is the reticular formation of the brainstem that comprises much of the brainstem core, known as the tegmentum. Although the ascending and descending reticular activating systems are well integrated, the latter tends to be centered in the medulla, whereas the former is found more in the pons and the midbrain. The reticular formation is essentially a heavily arranged network of neurons which are distributed throughout the brainstem, wherever there are no specific neural tracts or nuclei. The gigantocellular neurons receive a large portion of inputs from the spine. Pain has a strong influence on reticular activating system activity, particularly in the periaqueductal gray area of the midbrain. Many inputs to the reticular formation arise in higher brain centers, including the cerebellum, hypothalamus, basal ganglia, amygdala and the cerebral cortex (especially the premotor cortex) [13]. The gigantocellular neurons give rise to the ascending fibers which travel as the central tegmental tract to the intralaminar nuclei of the thalamus. There is positive feedback between an awake mind and the reticular activating system. The midbrain reticular formation sends projections to the hypothalamus, while paramedian reticular nuclei project mainly to the cerebellum. The parvicellular neurons receive inputs from the special senses further contributing to arousal.

Neurochemically speaking, the brainstem monoaminergic nuclei including noradrenergic, dopaminergic and serotonergic nuclei, project directly to the cortex, in order to modulate wakefulness and mood, as well as to the basal nuclei. Noradrenergic (locus coeruleus) and serotonergic (raphe) nuclei have been associated with attentiveness and behavioural responses to stimuli, while dopamine is, among others, associated with working memory by modulating excitatory and inhibitory transmission in the prefrontal cortex, and with motor control and reward. Moreover, although serotonergic neurons are silenced during REM sleep, they increase their activity during deep sleep. Central nervous system cholinergic nuclei include two major pathways: from the brainstem to he thalamus and from the brainstem, via the basal nuclei, to the cortex, thalamus, hippocampus and amygdala. Spiking activity in cholinergic neurons increases during wakefulness and REM sleep and decreases during non-REM or deep sleep. Acetylcholine is therefore believed to elevate arousal in the brain so as to allow the formation of consciousness, however fleeting. Histaminergic neurons of the posterior hypothalamus act like noradrenergic neurons in enforcing waking and are joined by neurons in the region that contain orexin, a neuropeptide recently shown to maintain waking and in absentia to be responsible for narcolepsy, or the inability to maintain wakefulness. These multiple arousal systems are grossly redundant, since no one system is absolutely necessary for the occurrence of waking; yet they are differentiated, since each plays a special role in waking and sleep. During SWS, they are submitted to an inhibitory influence arising in part at least from particular GABAergic neurons co-distributed with many neurons of the arousal systems and also concentrated within the basal forebrain and adjacent preoptic region [14, 15].

2. The Thalamus

Anatomically, the body of the thalamus is divided by a Y-shaped band of white matter, namely the internal medullary lamina, into three large cell groups: mediodorsal, anterior and lateral. Intralaminar nuclei are found within the lamina itself and a reticular nucleus surrounds the thalamus on the dorsal side. Functionally, the nuclei of the thalamus can be divided into three categories: relay nuclei, association nuclei and non-specific nuclei. The relay nuclei, whose function is to relay sensory and motor information to the cortex, include the geniculate bodies (hearing and vision stimuli) and the ventral nuclei (tactile and motor information). All these nuclei have reciprocal connections with the cortex, which definitely gives feedback-control on signals sent. The non-specific nuclei of the thalamus are the intralaminar nuclei, which are the site of termination for the ascending reticular activating system. The intralaminar neurons project cholinergic fibers diffusely to the cerebral cortex, providing generalized activation, and to the striatum. Central thalamic neurons appear to be specialized both anatomically and physiologically to support distributed network activity that maintains neuronal firing patterns across long-range cortico-cortical pathways and within cortico-striatopallidal-thalamo-cortical loop connections. Recruitment of central thalamic neurons occurs in response to increasing cognitive demand, stress, fatigue, and other perturbations that reduce behavioral performance. In addition, the central thalamus receives projections from brainstem pathways evolved to rapidly generate brief shifts of arousal associated with the appearance of salient stimuli across different sensory modalities. Through activation of the central thalamus, neurons across the cerebral cortex and striatum can be depolarized and their activity patterns selectively gated by descending or ascending signals related to premotor attention and alerting stimuli. The association nuclei are well-connected with other nuclei in the thalamus, and project largely to the association areas of the frontal and parietal lobes. In particular, the pulvinar seems to function for general integration of sensory information, with hearing and vision predominating, and to be important for the shifting of attention and suppression of irrelevant stimuli. Moreover, the thalamic reticular nucleus is actually only a thin sheet of inhibitor neurons that are believed to function as a "gate" for signals directed to the cerebrum. Thus, the reticular complex may control which sensory inputs are the subject of attention of the cerebral cortex [16].

Interestingly, the thalamic neurons sending projections to the cerebrum have specialized properties which enable them to fire in synchronized bursts or with desynchronized single-spike firing. The former corresponds to the EEG desynchronization seen in REM sleep or in the alert, awakened state. Recent theoretical advances describing consciousness from information and integration have highlighted the unique role of the thalamocortical system in leading to integrated information and thus, consciousness [17].

3. The Hypothalamus

The hypothalamus is the center of the autonomic nervous system and of the body's hormonal system and is important for the expression of emotion. The hypothalamus regulates body temperature, hunger, thirst, osmotic pressure, sex drive, etc.

Of the numerous and complex hypoyhalamic circuitries: 1) the suprachiasmatic nucleus receives fibers emerging from the optic nerve and seems to function in the regulation of circadian rhythms ("day-night cycles"); signals from the suprachiasmatic nucleus to the pineal gland results in the production of melatonin from serotonin. 2) the mammillary bodies in the posterior portion of the hypothalamus receive inputs mainly from the hippocampus via the fornix, and send outputs to the reticular formation and to the cingulate cortex via the mediodorsal thalamic nucleus; 3) the median forebrain bundle (MFB). The MFB is the main conduction pathway linking the brainstem (reticular formation, locus ceruleus, dorsal raphe and ventral tegmental area), hypothalamus and basal forebrain (amygdala and striatum). The hypothalamus, is also involved in the sleep-wake transition and its histaminergic outputs help maintain the awake state [15].

4. The Basal Forebrain

The basal forebrain is a term for a group of structures that lie near the bottom of the front of the brain, including the nucleus basalis (of Meynert), diagonal band, medial septum and substantia innominata. These structures are important in the production of a brain chemical called acetylcholine, which is then distributed widely throughout the brain. Acetylcholine affects the ability of brain cells to transmit information to one another, and also encourages plasticity, or learning. Thus, damage to the basal forebrain can reduce the amount of acetylcholine in the brain and impair learning. This may be one reason why basal forebrain damage can result in memory impairments such as amnesia and confabulation. Forebrain arousal is regulated by ascending influences from brainstem/basal forebrain neuronal populations ("arousal systems") and control signals descending from frontal cortical systems. These subcortical and cortical systems have converging projections to the central thalamus that emphasize their role in maintaining organized behavior during wakefulness [18,19].

5. Amygdala

The regions described as amygdala nuclei encompass several structures involved in many different functions, including emotional learning, memory modulation, aggression, social interactions, sexual orientation and arousal after specific stiluli. Among these nuclei are the basolateral complex, the cortical nucleus, the medial nucleus, and the central nucleus; the basolateral complex can be further subdivided into the lateral, the basal, and the accessory basal nuclei. The amygdala sends impulses to the hypothalamus for activation of the sympathetic nervous system, to the thalamic reticular nucleus for increased reflexes, to the nuclei of the trigeminal nerve and the facial nerve, and to the ventral tegmental area, locus coeruleus, and laterodorsal tegmental nucleus for activation of dopamine, norepinephrine and epinephrine. The cortical nucleus is involved in the sense of smell and pheromone-processing, as it receives input from the olfactory bulb and olfactory cortex. The lateral amygdalae, which send impulses to the rest of the basolateral complexes and to the centromedial nuclei, receive input from the sensory systems. The centromedial nuclei are the main outputs for the basolateral complexes, and are involved in emotional arousal. Emotional stimuli, including facial expressions, are thought to gain rapid and privileged access to processing resources in the brain.

Despite this access, we are conscious of only a fraction of the myriad of emotion-related cues we face every day. Considerable research supports the idea that contributions from the ventral visual system to consciousness vary in a category-selective manner [20,21]. However, to date little evidence exists that contributions from fronto-parietal areas considered critical for consciousness are also variable. A recent study by Amting et al. [22] have showed that activity in these regions can vary as a function of the emotion perceived and suppressed from awareness. Strikingly, although fearful stimuli were more likely to be perceived when present, this increased awareness was associated with reduced activity in regions of fronto-parietal cortices thought to play a critical and general role in consciousness. These results are consistent with the idea that the amygdala modulates the threshold for awareness of emotional stimuli. Evidence is also provided that activity in regions of pgPFC, previously implicated in extinguishing a fear-conditioned response and in explicit emotion regulation, is associated with the suppression of fearful faces from consciousness.

5. Basal Ganglia

The basal ganglia (BG) comprise two principal input nuclei, the striatum and the subthalamic nucleus (STN), and two principal output nuclei, the substantia nigra pars reticulata (SNr) and the internal globus pallidus (GPi), which in cats and rodents is known as the entopeduncular nucleus. The external globus pallidus (GPe) is principally an intrinsic structure that receives most of its afferents from, and provides efferent connections to other BG nuclei. Finally, dopaminergic neurones in substantia nigra (pars compacta) (SNc) and the adjacent ventral tegmental area (VTA) provide other BG nuclei, principally the striatum, with important modulatory signals.

Striatum is the largest nucleus of the BG. In primates the striatum comprises the caudate nucleus and the putamen, and in all mammals, the ventral striatum or nucleus accumbens. It receives direct input from most regions of the cerebral cortex and limbic structures including the amygdala and hippocampus. Additional input from sensorimotor and motivational regions of the brainstem arrives indirectly via relays in the thalamus. Finally, important modulatory afferents come from substantia nigra pars compacta (dopamine) and the raphe nuclei (serotonin) in the midbrain. Globus pallidus (internal) is one of the two output nuclei that receive inputs from other basal ganglia nuclei and provides output to external targets in the thalamus and brainstem. Globus pallidus (external) is the principal "intrinsic" structure of the basal ganglia since it's major connections are with other basal ganglia nuclei. Thus, it receives inhibitory input from the striatum, excitatory input from the subthalamus, and provides GABAergic inhibitory efferent connections to all the basal ganglia's input and output nuclei [23, 24].

Despite numerous suggestions that the basal ganglia are involved in a wide range of functions including perception, learning, memory, attention, many aspects of motor function, even analgesia and seizure suppression, increasingly evidence points to an underlying role in basic selection processes. In particular, the thalamus is a critical component of the frontal cortical-basal ganglia-thalamic circuits that mediate motivation and emotional drive, planning and cognition for the development and expression of goal-directed behaviors.

Each functional region of the frontal cortex is connected with specific areas of each basal ganglia (BG) structure and of the thalamus. In addition, the thalamus sends a massive, topographically organized projection directly to the striatum. Tract-tracing and physiological experiments have indicated a general topographic organization of the cortical-BG-thalamic loops and supported a model of BG function based on parallel and segregated pathways. However, the learning and execution of appropriate behavioral responses require integration of inputs related to emotional, cognitive, and motor cortical functions [25,16].

6. Prefrontal Cortex

The prefrontal cortex (PFC) is the anterior part of the frontal lobes of the brain, lying in front of the motor and premotor areas. This brain region has been implicated in planning complex cognitive behavior, personality expression, decision making and moderating social behaviour. The basic activity of this brain region is considered to be orchestration of thoughts and actions in accordance with internal goals.

The most typical psychological term for functions carried out by the prefrontal cortex area is executive function. Executive function relates to abilities to differentiate among conflicting thoughts, determine good and bad, better and best, same and different, future consequences of current activities, working toward a defined goal, prediction of outcomes, expectation based on actions, and social "control" (the ability to suppress urges that, if not suppressed, could lead to socially unacceptable outcomes). It has been indicated an integral link between a person's personality and the functions of the PFC [27].

Interestingly, a recent hypothesis sustains that the mental states commonly referred to as altered states of consciousness are principally due to transient prefrontal cortex deregulation. Supportive evidence from psychological and neuroscientific studies of dreaming, endurance running, meditation, daydreaming, hypnosis, and various drug induced states is presented and integrated. It is proposed that transient hypofrontality is the unifying feature of all altered states and that the phenomenological uniqueness of each state is the result of the differential viability of various frontal circuits [28].

7. The Cingulate Cortex

The cingulate cortex is a part of the brain situated in the medial aspect of the cortex. It includes the cortex of the cingulate gyrus, which lies immediately above the corpus callosum, and the continuation of this in the cingulate sulcus. The cingulate cortex is usually considered part of the limbic lobe. It receives inputs from the thalamus and the neocortex, and projects to the entorhinal cortex via the cingulum. It is an integral part of the limbic system, which is involved with emotion formation and processing, learning, and memory, and is also important for executive function. It has been recently demonstrated that the linkage of neural activity in the anterior cingulate cortex (ACC) during self-relatedness with the degree of consciousness patients with DOC. Most importantly, signal changes during self-relatedness in a caudal subregion of the ACC were correlated with the DOC patients'level of consciousness. This may also suggest a useful neural, and thus diagnostic, marker of the dysfunction of consciousness in vegetative patients [29].

8. Precuneus

The precuneus is a part of the superior parietal lobule hidden in the medial longitudinal fissure between the two cerebral hemispheres. It is sometimes described as the medial area of the superior parietal cortex. It is involved with episodic memory, visuospatial processing, reflections upon self, and aspects of consciousness. The location of the precuneus makes it difficult to study. Furthermore, it is rarely subject to isolated injury due to strokes, or trauma such as gunshot wounds. This has resulted in it being "one of the less accurately mapped areas of the whole cortical surface". Over the last few years, functional neuroimaging studies have started unravelling unexpected functional attributes for the precuneus. The posteromedial parietal areas are amongst the brain structures displaying the highest resting metabolic rates ("hot spots") and are characterized by transient decreases in the tonic activity during engagement in non self-referential goal-directed tasks ("default mode of brain function"). On the other hand, precuneus activation has been documented in healthy subjects engaged in self-related mental representation and episodic/autobiographical memory retrieval [30,31]. Moreover, selective precuneal hypometabolism has been reported in a wide range of altered conscious states, such as sleep, hypnosis, drug-induced anesthesia, vegetative state, and in neuropsychiatric conditions characterized by impaired consciousness (eg, Alzheimer's disease, epilepsy, schizophrenia). Taken together, these findings provide strong, albeit preliminary, evidence that this richly connected multimodal associative area belongs to the neural network subserving awareness and producing a conscious self-percept, a process that possibly runs in the background (by default) during silent rest. In particular, a major challenge for theorists of the default mode of brain functioning is to determine how task-induced deactivations (TIDs) and selective hypometabolism in pathophysiological conditions affecting consciousness relate to resting-state mentation.

One possibility is that when an individual is awake and alert and yet not actively engaged in particular cognitive task, the precuneus and interconnected posterior cingulate and medial prefrontal cortices subserve continuous information-gathering and representation of the self and external world [32]. This hypothesis fits nicely with the observed functional TIDs: when non-self-referential goal-directed processes are to be performed, the resting-state processes are interrupted, reflecting a necessary reduction in resources devoted to general information gathering and evaluation; as a consequence, it would seem to be a default activity of the brain with rather obvious evolutionary significance.

When the successful performance of a task demands focused attention, such broad information-gathering activity needs to be curtailed. Likewise, Binder et al. [33] suggested that precuneus activity during conscious resting states supports conceptual processing operating on internal stores of information (endogenous signals) rather than "perceptual" functions (concerned with sources of information external to the brain). In other words, this area seems to contribute to the self-referential "thought" processing that humans experience during resting consciousness.

Conclusion

Consciousness is dependent on the brain stem and thalamus for arousal, whereas basic cognition is supported by recurrent electrical activity between the cortex and the thalamus at gamma band frequencies; nevertheless, some kind of working memory, at least fleetingly, must be present for awareness to occur. It is generally accepted that the different features of objects are processed in different areas of the brain, and that the different qualities are cognitively bound together and experienced against the background of previous experiences - a process known as *cognitive binding*. However, for consciousness to be endowed with meaning that binds our perceptions into the seamless unity, - as suggested by Kant-, we require far-reaching explicit working memory which probably feeds on integrative circuits across the different domains of the prefrontal cortex.

Conversely, for the prefrontal structures to function properly in the generation of consciousness they must receive information from other brain areas such as other association cortices and the limbic system. Thus, consciousness also depends on the multitude of unconscious processes that occurs in the brain.

References

[1] Zeman A. What do we mean by "conscious" and "aware"? *Neuropsychol. Rehabil.* 2006;16:356-76.

[2] Seth AK, Baars BJ, Edelman DB. Criteria for consciousness in humans and other mammals. *Conscious. Cogn.* 2005;14;119-39.

[3] Young GB, Wijdicks EF. Consciousness: its neurological relevance. *Handb. Clin. Neurol.* 2008;90:33-6.

[4] Zeman A. Consciousness. *Brain.* 2001;124:1263-89.

[5] Moruzzi G, Magoun HW. Brain stem reticular formation and activation of the EEG. *Electroencephalogr. Clin. Neurophysiol.* 1949;1(4):455-73.

[6] Neylan TC. Physiology of arousal: Moruzzi and Magoun's ascending reticular activating system. *J. Neuropsychiatry Clin. Neurosci.* 1995;7(2):250.

[7] Laureys S. The neural correlate of (un)awareness: lessons from the vegetative state. *Trends Cogn. Sci.* 2005; 9(12):556-9.

[8] Palmiter RD. Dopamine signaling as a neural correlate of consciousness. *Neuroscience*. 2011;198:213-20.

[9] Negrao BL, Viljoen M. Neural correlates of consciousness. *Afr. J. Psychiatry* (Johannesbg). 2009;12(4):265-9.

[10] Heine L, Soddu A, Gómez F, Vanhaudenhuyse A, Tshibanda L, Thonnard M, Charland-Verville V, Kirsch M, Laureys S, Demertzi A. Resting state networks and consciousness: alterations of multiple resting state network connectivity in physiological, pharmacological, and pathological consciousness States. *Front Psychol*. 2012;3:295.

[11] Vanhaudenhuyse A, Demertzi A, Schabus M, Noirhomme Q, Bredart S, Boly M, Phillips C, Soddu A, Luxen A, Moonen G, Laureys S. Two distinct neuronal networks mediate the awareness of environment and of self. *J. Cogn. Neurosci*. 2011;23(3):570-8.

[12] Bob P, Zimmerman EM, Hamilton EA, Sheftel JG, Bajo SD, Raboch J, Golla M, Konopka LM. Conscious Attention, Meditation, and Bilateral Information Transfer. *Clin. EEG Neurosci*. 2012 Nov 19. [Epub ahead of print].

[13] Parvizi J, Damasio A. Consciousness and the brainstem. Cognition 2001;79(1-2):135-60.

[14] Jones BE. Arousal systems. *Front Biosci.* 2003;8:s438-51.

[15] Edlow BL, Takahashi E, Wu O, Benner T, Dai G, Bu L, Grant PE, Greer DM, Greenberg SM, Kinney HC, Folkerth RD. Neuroanatomic connectivity of the human ascending arousal system critical to consciousness and its disorders. *J. Neuropathol. Exp. Neurol*. 2012;71(6):531-46.

[16] Schiff ND. Central thalamic contributions to arousal regulation and neurological disorders of consciousness. *Ann. N. Y. Acad. Sci*. 2008;1129:105-18.

[17] Ward LM. The thalamic dynamic core theory of conscious experience. *Conscious. Cogn*. 2011; 20(2):464-86.

[18] Woolf NJ, Butcher LL. Cholinergic systems mediate action from movement to higher consciousness. *Behav. Brain Res*. 2011;221(2):488-98.

[19] Lucas-Meunier E, Fossier P, Baux G, Amar M. Cholinergic modulation of the cortical neuronal network. *Pflugers Arch*. 2003;446(1):17-29.

[20] Williams LM, Liddell BJ, Kemp AH, Bryant RA, Meares RA, Peduto AS, Gordon E. Amygdala-prefrontal dissociation of subliminal and supraliminal fear. *Hum. Brain Mapp*. 2006;27(8):652-61.

[21] Khetrapal N. The framework for disturbed affective consciousness in autism. *Neuropsychiatr. Dis. Treat.* 2008;4(3):531-3.
[22] Amting JM, Greening SG, Mitchell DG. Multiple mechanisms of consciousness: the neural correlates of emotional awareness. *J. Neurosci.* 2010;30(30):10039-47.
[23] McHaffie, J. G., Stanford, T. R., Stein, B. E., Coizet, W. and Redgrave, P. Subcortical loops through the basal ganglia. *Trends Neurosci.* 28, 401-407 (2005).
[24] Middleton, F. A. and Strick, P. L. Basal ganglia and cerebellar loops: motor and cognitive circuits. *Brain Res. Rev* 31, 236-250 (2000).
[25] Cotterill RM. Cooperation of the basal ganglia, cerebellum, sensory cerebrum and hippocampus: possible implications for cognition, consciousness, intelligence and creativity. *Prog. Neurobiol.* 2001 May;64(1):1-33.
[26] Haber SN, Calzavara R. The cortico-basal ganglia integrative network: the role of the thalamus. *Brain Res. Bull.* 2009;78(2-3):69-74.
[27] Edmund T. Rolls. Functions of the Orbitofrontal Cortex. *Brain and Cognition* 2004; 55: 11–29.
[28] Dietrich A. Functional neuroanatomy of altered states of consciousness: the transient hypofrontality hypothesis. *Conscious. Cogn.* 2003;12(2):231-56.
[29] Qin P, Di H, Liu Y, Yu S, Gong Q, Duncan N, Weng X, Laureys S, Northoff G. Anterior cingulate activity and the self in disorders of consciousness. *Hum. Brain Mapp.* 2010;31(12):1993-2002.
[30] Cavanna AE, Trimble MR. The precuneus: a review of its functional anatomy and behavioural correlates. *Brain.* 2006;129(pt 3):564-583.
[31] Cavanna AE. The precuneus and consciousness. *CNS Spectr.* 2007;12:545-52.
[32] Vogt BA, Laureys S. Posterior cingulate, precuneal and retrosplenial cortices: cytology and components of the neural network correlates of consciousness. *Prog. Brain Res.* 2005;150:205-17.
[33] Binder JR, Frost JA, Hammeke TA, Bellgowan PSF, Rao SM, Cox RW. Conceptual processing during the conscious resting state. A functional MRI study. *J. Cogn. Neurosci.* 1999;11:80-93.

In: Chronic Disorders of Consciousness
Editors: R. S. Calabrò and P. Bramanti
ISBN: 978-1-62618-940-9

Chapter 2

The Wide Spectrum of Consciousness Disorders

Rocco Salvatore Calabrò[1], Angela Marra[1], Giuseppina Digangi, Carmela Casella[2], Placido Bramanti[1] and Rosalia Silvestri[2]
[1]IRCCS Centro Neurolesi "Bonino-Pulejo", Messina, Italy
[2]Neurologic Unit, Policlinico Universitario, Messina, Italy

Abstract

Medically speaking, the term consciousness is the state of the patient's awareness of self and environment and his responsiveness to external stimulation and inner need. Temporary loss of consciousness may be caused by impaired cerebral perfusion (syncope, fainting), cerebral ischemia, migraine, epileptic seizures, metabolic disturbances, sudden increases in intracranial pressure, or sleep disorders. Chronic unconsciousness is a tragic and ironic failure of high-technology treatment to preserve or restore brain function, the primary aim of therapeutics. Management of a patient in a vegetative state or a minimally conscious state requires carefully reaching the correct diagnosis, pronouncing an evidence-based prognosis, and thoughtfully considering the medical, ethical, and legal elements of optimum treatment. This chapter is aimed at exploring the wide spectrum of consciousness disorders, with particular regards to those having a negative impact on quality of life of the patients and their care-givers, including epilepsy, parasomnias, coma and vegetative state.

Introduction

The terms *consciousness, confusion, stupor, unconsciousness*, and *coma* have been endowed with so many different meanings that it is almost impossible to avoid ambiguity in their use. Indeed, they are not only medical terms but literary, philosophic, and psychologic ones as well. The word *consciousness* is the most difficult of all.

William James once remarked that everyone knows what consciousness is until he attempts to define it. To the psychologist, consciousness denotes a state of continuous awareness of one's self and environment. Medically speaking, the term *consciousness* is the state of the patient's awareness of self and environment and his responsiveness to external stimulation and inner need. This narrow definition has an advantage in that *unconsciousness* has the opposite meaning -a state of unawareness of self and environment or a suspension of those mental activities by which people are made aware of themselves and their environment, coupled with a diminished responsiveness to environmental stimuli.

Normal Consciousness is the condition of the normal person when awake. In this state the individual is fully responsive to a thought or perception and indicates by his behavior and speech the same awareness of self and environment as that of the examiner [1].

Consciousness has 2 clinical components:

1. wakefulness, which is mediated by the ascending reticular activating system of the brainstem and its thalamic and hemispheric projections, and
2. awareness of one's self and the environment, which is mediated by the cerebral cortex and its connections with itself, the thalami, and other subcortical structures [2].

A disordered state of consciousness results when one or both of these components are compromised [3]. Disordered states of consciousness is a broad category that encompasses a spectrum of cognitive dysfunction, from mild confusional states, such as delirium, to dementia, coma, locked-in syndrome, vegetative state (VS), minimally conscious state (MCS), and brain death (figure 1).

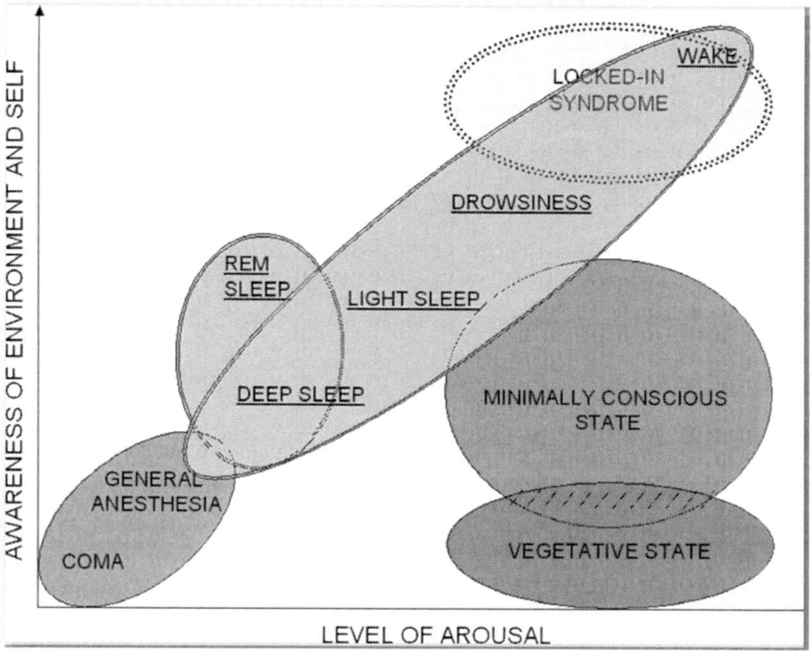

Figure 1. Shows the two major components of consciousness: wakefulness (i.e., the level of consciousness) and awareness of environment and self (i.e., the content of consciousness). In normal physiological states (underlined) arousal and awareness are positively correlated. Patients in l coma are unconscious because they cannot be awakened. The vegetative state illustrates the dissociation between arousal and awareness.

Episodic Impairment of Consciousness

Temporary loss of consciousness may be caused by impaired cerebral perfusion (syncope, fainting), cerebral ischemia, migraine, epileptic seizures, metabolic disturbances, sudden increases in intracranial pressure, or sleep disorders [4]. Anxiety attacks, psychogenic seizures, panic disorder, and malingering may be difficult to distinguish from these conditions. At times, the diagnosis may not be classified without detailed laboratory examinations and prolonged periods of observation.

Syncope may result from decreased cardiac output secondary to cardiac arrhythmias, outflow obstruction, hypovolemia, orthostatic hypotension, or decreased venous return. Temporary loss of consciousness may result in cerebrovascular disturbances from transient ischemic attacks of the posterior or anterior cerebral circulations or cerebral vasospasm from migraine, subarachnoid hemorrhage, or hypertensive encephalopathy. Metabolic disturbances caused by hypoxia, drugs, anemia, and hypoglycemia may cause in syncope or, more frequently, may present with the sensation of an impending faint (pre-syncope). Absence seizures, generalized tonic-clonic seizures, and complex partial seizures are associated with alterations of consciousness and, in most cases, can be easily distinguished from syncope. Epileptic seizures may be difficult to distinguish from pseudoseizures (psychogenic seizures), panic attacks, and malingering [5].

Epileptic Seizures

Seizures are the manifestation of abnormal hypersynchronous or hyperexcitable discharges of cortical neurons. The clinical signs or symptoms of seizures depend on the location of the epileptic discharges in the cerebral cortex and the extent and pattern of the propagation of the epileptic discharge in the brain. It should not be surprising that seizures are a common, non specific manifestation of neurologic injury and disease, because the main function of the brain is the transmission of electrical impulses. The lifetime likelihood of experiencing at least 1 epileptic seizure is about 9%, and the lifetime likelihood of receiving a diagnosis of epilepsy is almost 3%. However, the prevalence of active epilepsy is only about 0.8%. Indeed, epilepsy is defined as a brain disorder characterized by an enduring predisposition to generate epileptic seizures and by the neurobiologic, cognitive, psychologic, and social consequences of this condition. Traditionally, the diagnosis of epilepsy requires the occurrence of at least 2 unprovoked seizures 24 hours apart; however, some clinicians are also diagnosing epilepsy when one unprovoked seizure occurs in the setting of a predisposing cause such as a focal cortical injury or a generalized interictal discharge suggesting a persistent genetic predisposition. Seizures are divided into two categories: partial seizures (i.e., focal or localization-related seizures) and generalized seizures. *Partial seizures* result from a seizure discharge within a particular brain region or focus, and they manifest focal symptoms and may progress to secondarily generalized seizure, resulting in tonic-clonic activity. Some partial seizures have very rapid generalization, and the partial phase of the seizure may not be readily apparent clinically or even on scalp EEG recordings. However, secondarily generalized partial seizures are not included in the category of generalized seizures, which includes only primary generalized seizures. A complex partial seizure starts focally within the brain causing impairment of consciousness, and it is most commonly a manifestation of temporal lobe epilepsy, although the term is so broadly defined (i.e., as any focal seizure with impairment of consciousness) that it is very non-specific[6]. For this reason, many clinicians make a point of distinguishing between temporal and extratemporal complex partial seizures. *Primary generalized seizures* probably begin in the thalamus and other subcortical structures, but on scalp EEG recordings, they may appear to start simultaneously in both cerebral hemispheres; therefore, they manifest symptoms bilaterally in the body and are always associated with loss of consciousness. Generalized seizures can be classified as atonic, tonic, clonic, tonic-clonic, myoclonic, or absence on the basis of clinical symptoms and EEG abnormalities. Tonic seizure is the rigid contracture of muscles, including respiratory muscles, which is usually brief: the clonic component is the rhythmic shaking that occurs and is longer: together, a generalized tonic-clonic seizure (GTCS) is also called a grand mal seizure and is one of the most dramatic of all medical conditions. Patients with generalized tonic-clonic seizures and idiopathic generalized epilepsy typically have no evidence of any localized, regional, or diffuse brain abnormality on history, physical, or neurologic examination; clinical laboratory testing; or imaging studies. The awake EEG of patients with generalized tonic-clonic seizure may be normal; however, certain specific interictal EEG patterns can be distinctive of generalized epilepsy syndromes [7].

Status epilepticus (SE) is a life-threatening condition in which the brain is in a state of persistent seizure. Definitions vary, but traditionally it is defined as one continuous, unremitting seizure lasting longer than 30 minutes, or recurrent seizures without regaining consciousness between seizures for greater than 30 minutes. Treatment is, however, generally started after the seizure has lasted five minutes. It is always considered a medical emergency. There is some evidence that five minutes is sufficient to damage neurons and that seizures are unlikely to self-terminate by that time. First aid guidelines for seizures state that, as a rule, an ambulance should be called for seizures lasting longer than five minutes (if this is the patient's first seizure episode and no precipitating factors are known, or if SE happens to an epileptic whose seizures were previously absent or well controlled for a considerable time period, then that step can be taken before that point). The mortality rate of status epilepticus has the potential to be quite high (at least 20%), especially if treatment is not initiated quickly. However, with optimal neurological care, adherence to the medication regimen, and a good prognosis (no other underlying uncontrolled brain or other organic disease), the patient in otherwise good health can survive with minimal or no brain damage, and can decrease their risk of death and even avoid future seizures [8].

Neurophysiology of Seizures

Physiologically, the epileptic seizure has been defined as a sudden alteration of central nervous system (CNS) function resulting from a paroxysmal high-frequency or synchronous low-frequency, high voltage electrical discharge. This discharge arises from an assemblage of excitable neurons in any part of the cerebral cortex and probably in secondarily involved subcortical structures as well. In the proper circumstances, a seizure discharge can be initiated in an entirely normal cerebral cortex, as when the cortex is activated by ingestion or injection of drugs, by withdrawal from alcohol or other sedative drugs, or by repeated stimulation with subconvulsive electrical pulses ("kindling phenomenon") [9].

Seizures require three conditions:

- a population of pathologically excitable neurons;
- an increase in excitatory glutaminergic activity through recurrent connections in order to spread the discharge;
- a reduction in the activity of the normally inhibitory GABA-nergic projections.

Understanding of the initial discharges and their spread has been greatly advanced by the identification of several rare forms of familial epilepsy that are the direct result of mutations in sodium, potassium, acetylcholine receptor, or GABA channels on neurons [10].

To date, the reason why the neurons in or near a focal cortical lesion discharge abnormally is not fully understood, albeit some of the electrical properties of a cortical epileptogenic focus suggest that its neurons have been deafferented. Such neurons are known to be hyperexcitable, and they may remain so chronically, in a state of partial depolarization, able to fire irregularly at rates as high as 700 to 1000 per second. The cytoplasmic membranes of such cells appear to have an increased ionic permeability, which renders them susceptible to activation by hyperthermia, hypoxia, hypoglycemia, hypocalcemia, and hyponatremia as well as by repeated sensory (e.g., photic) stimulation and during certain phases of sleep (where *hypersynchrony* of neurons is known to occur).

However, The precise mechanisms that govern the transition from a circumscribed interictal discharge to a widespread seizure state are not understood. Recent data indicate that the characteristic EEG patterns of both generalized forms of epilepsy (i.e., tonic or tonic clonic and absence) are generated in the neocortex and are enhanced by the synchronizing influences of subcortical structures [11].

In both instances, the generalization of the clinical and electrical manifestations depends upon activation of a deep, centrally located physiologic mechanism, which includes the midbrain reticular formation and its diencephalic extension, the intralaminar and nonspecific thalamic projection systems (originally referred to by Penfield as the "centrencephalon") [12].

There is no evidence, however, that seizure activity originates in these deep activating structures; therefore the term *centrencephalic epilepsy* has been replaced by *corticoreticular epilepsy*.

Complex partial seizures are almost always of temporal lobe origin, arising in foci in the medial temporal lobe, amygdaloid nuclei, and hippocampus. Nenertheless, they rarely originate in the convexity of the temporal lobe and propagate to the amygdaloid nuclei, hippocampus, and posteroinferior parts of the frontal lobe. Electrical stimulation in these areas reproduces feelings of depersonalization, emotionality, and automatic behavior, the characteristic features of psychomotor epilepsy. The automatic behavior appears to be a direct effect of the temporal lobe discharge in some instances and a postexcitatory or inhibitory effect in others. Loss of memory for the events of the episode may be due to the paralytic effect of the discharge on neurons of the hippocampus. Of theoretical importance is the observation that a seizure focus, if active for a time, may sometimes establish, via commissural connections, a persistent secondary focus in the corresponding cortical area of the opposite hemisphere (*mirror focus*).

The nature of this phenomenon is obscure; it may be similar to the "kindling"phenomenon mentioned earlier in animals, where a repeated non-convulsive electrical stimulation of normal cortex induces a permanent epileptic focus.

Psychogenic nonepileptic seizures (PNES), or pseudoseizures are paroxysmal episodes that resemble and often misdiagnosed as epileptic seizures; however, PNES are psychological (ie, emotional, stress-related) in origin. Patients with PNES have a higher prevalence of past traumatic experiences, PTSD, anxiety, depression, epilepsy, fibromyalgia and chronic pain. They often also convey their history differently than patients with epilepsy do; they may avoid reporting seizure symptoms or present the gaps in consciousness as the most dominant complaint, and the interviewer may have to suggest seizure as a possibility. PNES are common at epilepsy centers, where they are seen in 20-30% of patients referred for refractory seizures. PNES are probably also common in the general population, with an estimated prevalence of 2-33 cases per 100,000 population, which makes PNES nearly as prevalent as multiple sclerosis or trigeminal neuralgia [13].

Confusion. The term *confusion* lacks precision, but in general it denotes an inability to think with customary speed, clarity, and coherence. All states of confusion are marked by some degree of inattentiveness and disorientation. In this condition the patient does not take into account all elements of his immediate environment, and it also implies a degree of imperceptiveness and distractibility, referred to traditionally as "clouding of the sensorium. Confusion results most often from a process that influences the brain globally, such as a toxic or metabolic disturbance or a dementia. However, a confusional state can also accompany focal cerebral disease in various locations, particularly in the right hemisphere, or result from disorders that may disturb mainly language, memory, or visuo-spatial orientation, but these are readily distinguished from the global confusional state.

Delirium. In some medical writings, particularly in the psychiatric literature, the terms delirium and confusion are used interchangeably, the former connoting nothing more than a nondescript confusional state in which hyperactivity may be prominent. However, in the syndrome of delirium tremens (observed most often but not exclusively in alcoholics), the vivid hallucinations, the inaccessibility of the patient to events other than those to which he is reacting at the moment, the extreme agitation, the tendency to tremble, startle easily and the convulse, and the signs of overactivity of the autonomic nervous system, as well as a normal EEG tracing, suggest that the term *delirium* should be reserved for a highly distinctive confusional syndrome [14].

Drowsiness and Stupor. In these states, mental, speech, and physical activity are reduced. *Drowsiness* denotes an inability to sustain a wakeful state without the application of external stimuli. Inattentiveness and mild confusion are the rule, both improving with arousal: the patient shifts positions in bed or chair, the lids droop without closing completely; there may be snoring, the jaw and limb muscles are slack, and the limbs are relaxed. This state is indistinguishable from light sleep, with slow arousal elicited by speaking to the patient or applying a tactile stimulus.

Stupor describes a state in which the patient can be roused only by vigorous and repeated stimuli, at which time he opens his eyes, looks at the examiner, and does not appear to be unconscious; response to spoken commands is either absent or slow and inadequate; restless or stereotyped motor activity is common in stuporous patients and there is a reduction in the natural shifting of positions. When left unstimulated, these patients quickly drift back into a sleep-like state.

Coma

The most common disorder of consciousness that immediately follows severe brain injury is coma. Coma is a state characterized by eyes-closed unresponsiveness: comatose patients fail to respond to even the most vigorous stimulation, and when given noxious stimulation, patients may not move at all or may display stereotyped/reflexive movements only. Coma is always a symptomatic expression of an underlying disease. Sometimes the primary disorder is perfectly obvious, as with severe cranial trauma. All too often, however, the patient is brought to the hospital in a state of coma and little pertinent medical information is immediately available.

When the comatose patient is first seen, one must quickly make sure that the patient's airway is clear and blood pressure is adequate; if trauma has occurred, one must check for bleeding from a wound or ruptured organ (e.g., spleen or liver). When hypotension is present, certain therapeutic measures-placement of a central venous line and administration of fluids and pressor agents, oxygen, blood, or glucose solutions (preferably after blood is drawn for glucose determinations and thiamine is administered)-take precedence over diagnostic procedures. If respirations are shallow or labored or if there is emesis with a threat of aspiration, tracheal intubation and mechanical ventilation are required. The patient with a head injury may also have suffered a fracture of the cervical vertebrae, in which case caution must be exercised in moving the head and neck as well as in intubation, lest the spinal cord be inadvertently damaged. There must then be an inquiry as to the previous health of the patient-whether there was a history of diabetes, a head injury, a convulsion, alcohol or drug use, or a prior episode of coma or attempted suicide—and the circumstances in which the person was found. Persons who accompany the comatose patient to the hospital should be encouraged not to leave until they have been questioned [15].

When one is assessing stupor and coma that has developed in a hospitalized patient who does not display localizing findings, it is instructive to review the patient's medications closely. A large category of compounds may reduce alertness to the point of profound somnolence or stupor, particularly if there are underlying medical problems (especially liver failure).

Outstanding in lists of iatrogenic drug intoxications are anaesthetics, sedatives, anticonvulsants, opiates, antidepressants, and some antipsychotic compounds. From an initial survey, many of the common causes of coma-such as severe head injury, alcoholic or other forms of drug intoxication, and hypertensive brain hemorrhage-are readily recognized.

Prognosis. As a general rule, recovery from coma of metabolic and toxic causes is far better than from anoxic coma, with head injury occupying an intermediate prognostic position. Most patients who are comatose as a result of a stroke will die; subarachnoid hemorrhage in which coma is due to hydrocephalus is an exception. Postanoxic coma is a state of unconsciousness caused by global anoxia of the brain, most commonly due to cardiac arrest. Outcome after postanoxic coma lasting more than several hours is generally, but not invariably, poor [17]. Recovery of consciousness reported in the literature varies from 8% to 72% of patients, but is mostly thought to be around 20-30% in patients surviving in coma for at least 24 h. Research is directed at defining factors that reliably predict poor outcome in these patients. Favorable outcome proves impossible to predict. Studies on outcome prediction have focused mostly on neurological examination, clinical neurophysiological tests and biochemical parameters. The most recent and extensive study in this respect was the PROPAC study in The Netherlands [17], which confirmed earlier findings that bilaterally absent early cortical response after median nerve somatosensory potentials (absent somatosensory evoked potentials) is the most reliable predictor of poor outcome (no recovery of consciousness). A serum neuron-specific-enolase level >33 microg L(-1) seemed equally reliable (ref). In regard to all forms of coma, but particularly after cardiac arrest, if there are no pupillary, corneal, or oculo-vestibular responses within several hours of the onset of coma, the chances of regaining independent function are practically nil. Other signs that predict a poor outcome are absence of corneal reflexes and eye opening responses and atonia of the limbs at 1 and 3 days after the onset of coma and absence of the cortical component of the somatosensory evoked responses on both sides. The survivor from this latter group may unfortunately remain in a vegetative state for months or years, breathing without aid and with preserved hypothalamo-pituitary functions.

Relationship of Sleep to Coma Persons in sleep give little evidence of being aware of themselves or their environment; in this respect they are unconscious.

Sleep shares a number of other features with the pathologic states of drowsiness, stupor, and coma. These include yawning, closure of the eyelids, cessation of blinking and swallowing, upward deviation or divergence or roving movements of the eyes, loss of muscular tone, decrease or loss of tendon reflexes, and even the presence of Babinski signs and irregular respirations, sometimes Cheyne-Stokes in type. Upon being awakened from deep sleep, a normal person maybe confused for a few moments, as every physician knows. Nevertheless, sleeping persons may still respond to unaccustomed stimuli and at times are capable of some mental activity in the form of dreams that leave traces of memory, thus differing from persons in stupor or coma.

The most important difference, however, is that persons in sleep, when stimulated, can be roused to normal consciousness. There are important physiologic differences as well: cerebral oxygen uptake does not decrease during sleep, as it usually does in coma and recordable electrical activity and spontaneous motor activity differ in the two states, as indicated later in this chapter.

Brain Death

In the late 1950s European neurologists paid attention to a state of coma in which the brain was irreversibly damaged and had ceased to function but pulmonary and cardiac function could still be maintained by artificial means. Mollaret and Goulon referred to this condition as *coma de´passe´* (a state beyond coma).

A Harvard Medical School committee, in 1968, called it *brain death* and established a set of clinical criteria by which it could be recognized [18]. RD Adams, who was a member of the committee, defined the state as one of complete unresponsiveness to all modes of stimulation, respiratory arrest, and absence of all EEG activity for 24 h. The concept that a person is dead if the brain is dead and that death of the brain may precede the cessation of cardiac function has posed a number of important ethical, legal, and social problems as well as medical ones.

The central considerations in the diagnosis of brain death are:

1. absence of cerebral functions;
2. absence of brainstem functions, including spontaneous respiration;
3. irreversibility of the state. Following from the last criterion, it is necessary to demonstrate an irrefutable cause of the catastrophic brain damage (i.e., trauma, cardiac arrest, cerebral hemorrhage) and to exclude reversible causes such as drug overdose [19].

The absence of cerebral function is judged by the presence of deep coma and total lack of spontaneous movement and of motor and vocal responses to all visual, auditory, and cutaneous stimulation. Spinal reflexes may be present in some cases, and the toes often flex slowly in response to plantar stimulation; but a well-developed Babinski sign is unusual in our experience (although its presence does not exclude the diagnosis of brain death). Extensor or flexor posturing is seen from time to time as a transitional phenomenon just after brain death becomes evident. The absence of brainstem function is judged by the loss of spontaneous eye movements, midposition of the eyes, and lack of response to oculocephalic and caloric (oculovestibular) testing; presence of dilated or midposition fixed pupils (not smaller than 3 mm); paralysis of bulbar musculature (no facial movement or gag, cough, corneal, or sucking reflexes); an absence of motor and autonomic responses to noxious stimuli; and absence of respiratory movements. As a final test of this last component, it has become customary to perform an "*apnea test*" to demonstrate an unresponsivity of the medullary centers to a high carbon dioxide tension. This test is conducted by first employing preoxygenation for several minutes with 90% inspired oxygen, the purpose of which is to displace nitrogen from the alveoli and create a reservoir of oxygen that will diffuse down a gradient into the pulmonary blood. The patient can then be disconnected from the respirator for a few minutes (during which time 100% oxygen is being delivered by cannula); this allows the PCO to rise to 50 to 60 mmHg (typically, carbon dioxide 2 rises 2.5 mmHg per minute at normal body temperature —slower if the patient is hypothermic). If no breathing is observed and examination of the blood gases shows that an adequate level of PCO has been attained, the presence of this component of brain death is corroborated. The EEG provides confirmation of cerebral death, and most institutions prefer corroboration of the clinical features by the demonstration of electrocerebral silence ("flat" or, more accurately, isoelectric EEG, shown first by Schwab), which is considered to be present if there is no electrical potential of more than 2 mV during a 30-min recording (except for artifacts created by the ventilator, ECG, or surrounding electrical devices). There are cases on record in which a patient with an isoelectric EEG has had preserved brainstem reflexes [20]. It must also be emphasized that cerebral unresponsivity and a flat EEG do not always signify brain death; both may occur and may be reversible in states of profound hypothermia or intoxication with sedative-hypnotic drugs and immediately following cardiac arrest. For this reason, it has been recommended that the diagnosis of brain death not be entertained until several hours have passed from the time of initial observation. If the examination is performed at least 6 h after the ictus and there is evidence of overwhelming brain injury from trauma, anoxia, or massive cerebral hemorrhage (the most common conditions causing brain death), there is no need for serial testing. If cardiac arrest was the antecedent event, or the cause of neurologic damage is unclear, or drug or alcohol intoxication could reasonably have played a role in suppressing the brainstem's reflexes, it is advisable to wait about 24 h before pronouncing the patient dead. Toxicologic screening of the serum or urine is requisite in the latter circumstances. Evoked potentials show interesting but variable abnormalities in brain-dead patients but are not of primary value in the diagnosis. Some centers use nuclide brain scanning or cerebral angiography to demonstrate an absence of blood flow to the brain, equating this with brain death; but there are technical pitfalls in the use of these methods, and it is preferable to keep the diagnosis of death primarilyclinical. The same can be said for transcranial Doppler sonography, which in brain death shows a to-and-fro "*pendelfluss*" blood-flow pattern in the basal vessels [19, 20]. These tasks often fall to the neurologist. From our experience, it is best not to embark on clinical or EEG testing for brain death unless there is a clear intention on the part of the physician to remove the ventilator or request organ donation at the end of the process. The process and its intended outcome should be explained to the family. The family's desires regarding organ transplantation should be sought after adequate time has passed for them to absorb the shock of the circumstances. Neurologists must, of course, resist pressures from diverse sources that might lead them to the premature designation of a state of brain death. At the same time, it should be clarified that while brain death is an operational state that allows transplantation to proceed and mandates withdrawal of ventilation and blood pressure support, patients with overwhelming brain injuries need not fulfill these absolute criteria in order for medical support to be withdrawn.

A task force for the determination of *brain death in children* has recommended the adoption of essentially the same criteria as for adults. Because of the great difficulty in evaluating the status of nervous function in relation to perinatal insults, they have suggested that a diagnosis of brain death not be made before the seventh postnatal day and that the period of observation be extended to 48 h. As with adults, the possibility of reversible brain dysfunction from toxins, drugs, hypothermia, and hypotension must always be considered [21].

Chronic Unconsciousness

This is a tragic and ironic failure of high-technology treatment to preserve or restore brain function, the primary aim of therapeutics. Management of a patient in a VS or a minimally conscious state (MCS) requires carefully reaching the correct diagnosis, pronouncing an evidence-based prognosis, and thoughtfully considering the medical, ethical, and legal elements of optimum treatment [22].

Vegetative State and Minimally Conscious State

The vegetative state and the minimally conscious state are disorders of consciousness that can be acute and reversible or chronic and irreversible. Diffuse lesions of the thalami, cortical neurons, or the white-matter tracts that connect them cause the vegetative state, which is wakefulness without awareness. Functional imaging with PET and functional MRI shows activation of primary cortical areas with stimulation, but not of secondary areas or distributed neural networks that would indicate awareness [23, 24].

VS is a behaviorally-defined state, whereby patients show no evidence of self or environmental awareness. As in coma state patients can have spontaneous or stimulus-induced, stereotyped movements, and may retain brainstem regulation of visceral autonomic function that would suggest that the lower brainstem is intact [25]. However, the main behaviorally salient difference from coma is that VS patients cycle daily through eyes-open and eyes closed periods. This does not imply that VS patients have normal sleep-wake cycles; rather, their EEGs display a monotonous slow pattern regardless of whether the eyes are open or closed, or they only have fragmented components of normal electrographic sleep-wake phenomenology. The periods of eye opening reflect only a crude arousal pattern that involves upper brainstem nuclei. VS may represent a transitional state on the way to recovery of consciousness or could be a chronic condition in cases of more severe brain injuries. *Persistent vegetative state* (PVS) is a term used for patients who have remained in VS for an arbitrarily defined duration of 30 days. Another commonly used term is *Permanent vegetative* state [26], which is applied to patients in VS after global ischemia for 3 months or TBI for 1 year. Permanent VS, is more of a prognosis than a diagnosis, as these time durations reflect only a reduced probability of recovery. The next level of recovery on the continuum from VS to full consciousness is the MCS, defined as "a condition of severely altered consciousness in which minimal but definite behavioral evidence of self or environmental awareness is demonstrated." MCS includes a more heterogeneous group of patients than VS, because the operational definition allows for a wide range of behaviors, whereas VS only includes reflexive movements. In MCS, low-end behaviours include visual tracking to a mirror, localization of noxious touch, and inaccurate verbalization, whereas high-end behaviors include consistent movement to command and choosing correctly between 2 objects [27]. Patients with only low-end behaviors can be difficult to differentiate from VS, because these behaviors may be subtle and infrequent. This differentiation is essential because patients in MCS have significantly better prognoses for recovery than those in VS [28].

The term "apallic", in the Apallic Syndrome (see the Gerstenbrand Renission Scale), cannot be explained by or taken nowadays for a conditio sine qua non of an anatomically completed and permanent disconnection of neocortical structures and of higher cerebral functioning from the midbrain when, awaking from coma, functional remission is possible to a certain stage as opposed to the individuals suffering from final end stage of an incurable progressive neurological disease. In so far "apallic" is functional-anatomically incorrect, but clinically still appropriate for neurobehavioral symptoms, signs, and functional remission. Bycontrast, persistent "vegetative" state (PVS) must not necessarily be "persistent", why Jennett changed the phraseto "permanent", well aware that one has to accept that the abbreviation "P" will still widely be used to mean persistent rather than permanent in a non-acceptable misleading way, albeit a persisting negative connotation of "vegetative". The name "vegetative" was originally chosen to refer to the preserved vegetative (autonomous) nervous functioning of preserved sleep-wake cycles, respiration, digestion and thermoregulation.

Apallic Syndrome – Modified Innsbruck Remission Scale – 8 Phases (Gerstenbrand, 2011)

Phase 1	Deep somnolent, temporary open eyes, optical fixation, sleep-wake rhythm fatigue regulated, primitive emotional reactions, primitive motor patterns partly diminished, flexed-stretched extremity position, remaining mass movements, rigido-spasticity.
Phase 2	Somnolent, optical tracking, sleep-wake rhythm begin of day time regulation, emotional reaction tendency to differentiation, primitive motor patterns tendency aim directed, diminished mass movements, tendency to finalizing, diminishing of flexed-stretched body position, diminished of rigido-spasticity.
Phase 3	Begin of responsible wakefulness, somnolence phases, following simple commands, emotional reaction differentiated (positive, negative), primitive motor patterns differentiated (higher organized grasping reflexes, oral reflexes), oral feeding accepted, first finalized movements initiated by commands.
Phase 4	Klüver-Bucy-Phase, wakeful, sleep-wake rhythm daytime regulated, object grasping with attempt to bite and chew, no recognition of the objects, increased interest for the genital region, reacting to simple orders, producing primitive sounds, beginning signs of local brain lesions, rest of flexed-stretched body position, rest of rigido-spasticity.
Phase 5	Post Klüver-Bucy-Phase, wakeful, sleep-wake rhythm day-night adapted, slight rest of flexed-stretched body position, slight signs of spasticity, diminishing of the Klüver-Bucy-patterns, increasing of finalized movements, production of simple words, upcoming of local brain lesion signs.
Phase 6	Phase of Korsakow-Symptoms, fully awake, disorientation, fully guidable fatigue phases, rest of primitive motor patterns, rest of spasticity, directed finalized movements, guidable, begin of walking.
Phase 7	Phase of the amnestic syndrome, fully awake, day-night regulated sleep-wake rhythm, rest of disorientation, fully guidable, severe memory disturbances, rest of primitive motor patterns, slight signs of spasticity, prompt finalized movements, marked diffuse and local cerebral lesions possible, symptoms of Bed Rest Syndrome.
Phase 8	Phase of the end of remission state; begin of defect state, undisturbed consciousness, normal sleep-wake rhythm, increased fatigue phases, marked symptoms of diffuse and/or local cerebral lesions (neurological, cognitive behavior deficits) possible, Bed Rest symptoms, remarks of different handicaps.

Given these concerns regarding the negative connotation inherent in the term *vegetative state* and its possible effect on vulnerable patients awakening from coma, who sometimes never recover any voluntary responsiveness but may (probably more often than initially believed) recover minimal signs of consciousness, it has been proposed to change the label *vegetative state*, thus hoping to make it easier to change their management and standards of care. The European Task Force on Disorders of Consciousness has passed a proposal to change the name to *unresponsive wakefulness syndrome* or UWS [30]. It should be stressed that UWS is a clinical syndrome describing patients who fail to show voluntary motor responsiveness in the presence of eyes-open wakefulness which can be either transitory on the way to recovery from (minimal) consciousness or irreversible.

Akinetic mutism, in its originally described form, fits in the category of MCS, although milder variants, including abulia, are categorized as fully conscious states. Patients with these conditions have intact arousal and often the appearance of vigilance, but have severe poverty of movement despite lack of damage to motor systems. Severe cases can only be distinguished from VS by the preservation of visual tracking through smooth pursuit eye movements. The underlying cause in most cases is injury to the bilateral medial frontal lobes and anterior cingulate cortex from a mass lesion or anterior cerebral artery infarct. These syndromes can also arise from bilateral injury to the basal ganglia, dorsal and central thalamus, or midbrain, as these areas are tightly integrated with the frontal lobes in the generation of goal-directed behaviour [31].

Locked-in Syndrome

Locked-in syndrome is a condition in which a patient is aware and awake but cannot move or communicate verbally due to complete paralysis of nearly all voluntary muscles in the body except for the eyes. Albeit this is not a consciousness disorders, it should be briefly discussed since it may pose some problem in differential diagnosis.

Locked-in syndrome usually results in quadriplegia and the inability to speak in otherwise cognitively intact individuals. Those with locked-in syndrome may be able to communicate with others through coded messages by blinking or moving their eyes, which are often not affected by the paralysis. The symptoms are similar to those of sleep paralysis. Patients who have locked-in syndrome are conscious and aware, with no loss of cognitive function. They can sometimes retain proprioception and sensation throughout their bodies. Some patients may have the ability to move certain facial muscles, and most often some or all of the extraocular eye muscles. Individuals with the syndrome lack coordination between breathing and voice. This restricts them from producing voluntary sounds, though the vocal cords are not paralysed [32].

Dementia

Beside VS and MCS, dementia is a different form of chronic unconsciousness condition that should be briefly considered herein. Dementia (taken from Latin, originally meaning "madness", from *de-* "without" + *ment*, the root of *mens* "mind") is a serious loss of global cognitive ability in a previously unimpaired person, beyond what might be expected from normal aging. Dementia is not a single disease, but rather a non-specific illness syndrome (i.e., set of signs and symptoms) in which affected areas of cognition may be memory, attention, language, and problem solving. It is normally required to be present for at least 6 months to be diagnosed; cognitive dysfunction that has been seen only over shorter times, in particular less than weeks, must be termed delirium. In all types of general cognitive dysfunction, higher mental functions are affected first in the process [33]. Dementia which begins gradually and worsens progressively over several years is usually caused by neurodegenerative disease; i.e. by conditions affecting only or primarily the neurons of the brain and causing gradual but irreversible loss of function of these cells. Less commonly, a non-degenerative condition may have secondary effects on brain cells, which may or may not be reversible if the condition is treated. The causes of dementia depend on the age at which symptoms begin. In the elderly population (usually defined in this context as over 65 years of age), a large majority of cases of dementia are caused by Alzheimer's disease, vascular dementia or both.

Sleep as a Model for Altered States of Consciousness

Human and primate sleep is subdivided into NREM and REM sleep. During NREM sleep there is a progressive slowing of brain metabolic activity and flow from wakefulness, most prominent over the anterior part of the cortex, in the dorsolateral prefrontal cortex (DLPC) responsible for the executive functions. This is paralleled by progressive slowing and synchronization of electrical brain activity which becomes slower in frequency with increasingly high voltage. Serotonin and noradrenalin neurotransmission slow down, while histaminergic and gabaergic modulation take over. Motor activity is dysfacilitated occurring only on sleep stage transitions. Mental activity becomes perseverative and akin to rumination; perception of external stimuli is dull or absent.

Instead, during REM sleep, cholinergic firing from the latero-dorsal tegmentum (LDT) and the peduncolo pontine nucleus (PPN), stimulate the occipital cortex inducing vivid visual imagery which is internally generated due to complete shut-off of output-input gating. Mental activity is discontinuous, bizarre and non-logical, given the failure of memory and attention sustained by the shutdown of noradrenergic and serotoninergic transmission. Movement is not only dysfacilitated as in NREM sleep, but actually actively inhibited at the level of spinal motor neurons through the sublateral dorsal nucleus (SLD) stimulation.

Besides this selective inhibition, REM sleep carries a cumulative excitation level similar to that observed during the waking state, with the exception of the fronto-parietal cortices which still maintain the low level of activity of slow wave sleep (SWS) and are therefore unable to monitor and critically supervise the vivid internally generated brain constructs. This uninhibited excitatory state may be thus viewed as a kind of ancestral protoconsciousness where all virtual motor and memory programs of waking may be tried and played out in a fictive inner space, separated from wake experiences [34].

Sleep with it's different stages, three for NREM sleep and REM sleep, should not however be viewed as a unitary concept.

Recent animal experiments [35] demonstrated that some brain areas may be asleep while the rest of the brain is still awake. In particular, neurons can go offline during prolonged wake. Off periods in the frontal motor cortex increase as a function of sleep pressure following sleep deprivation. During these periods of "local sleep" rats appear awake and active, but become progressively impaired in motor task reactions.

It is unclear whether local sleep in certain animal species may be considered as an adaptive or maladaptive response. Also, the ability to monitor the environment while asleep in certain animal species could be evolutionarily advantageous to achieve better survival prospects against predatory assaults.

Neurosurgically implanted humans with hippocampal electrodes [36] showed similar behavior, demonstrating the occurrence of hippocampal spindling preceding by several minutes the occurrence of neocortical sleep onset on the scalp. The automatic amnestic behavior typical of known sleep disorders, such as sleep apnea and narcolepsy, could be due to local phenomena of sleep-wake dissociation. In other words, Wake, NREM and REM sleep are not to be seen as discreet all-or-nothing states of being, but rather their intrinsic components may rapidly oscillate and recombine in hybrid states of pathological (sleep disorders) or simply unusual (parasomnias) consciousness misfits [37]. Some clinical examples of dissociated states of consciousness mediated from sleep pathology include narcolepsy, disorders of arousal (DOA) and REM related parasomnias.

Narcolepsy

Narcolepsy is a rare neurological sleep disorder characterized by a tetrad of symptoms: excessive daytime sleepiness (EDS) with irresistible nap urgency, cataplexy, hypnagogic/hypnopompic hallucinations and paralysis. All of the latter denounce the rapid oscillation of sleep components such as loss of muscle tone and/or dream-like imagery during wakefulness or during the transition between REM sleep and the waking state. The lack of hypothalamic orexin/hypocretin in this disease is responsible for both sleepiness and the incorrect coupling of sleep markers across sleep-wake phases. An exaggerated REM propensity explains the persistence of REM sleep markers at the sleep-wake transition (hypnagogic hallucinations and paralysis) or during wakefulness through the inhibition of muscle tone, prompted by strong emotions, during cataplectic attacks. Sleep onset REM periods (SOREMPs) both at night and during daytime naps are the distinctive physiological markers of the disease, even if in specific conditions, such as severe sleep deprivation by drugs, environmental or clinical conditions, SOREMPs may be found less frequently and consistently than in narcolepsy. REM suppressant drugs such as SSRIs or other antidepressants, sodium oxybate – a gaba-b agonist – and modafinil are indicated as the current treatment options for narcolepsy [38].

Disorders of Arousal

Disorders of arousal (DOA) are NREM parasomnias, undesirable, unusual phenomena occurring as behavioral, autonomic or experiential events. They occur out of deep stages of NREM, typically during SWS in the first third of the night. They are partially due to an inherent predisposition toward arousal instability, tend to recur as a familial trait, precipitated by sleep deprivation alone or due to other sleep disorders responsible for sleep instability. They are also called "disorders of partial arousal" since the subject is basically unable to fully awaken and manifest instinctual behaviors typical of the waking state during apparent sleep. DOA include sleep walking, night terrors and confusional arousals.

All of the latter are by definition altered states of consciousness to be distinguished in a differential diagnosis from other paroxysmal events such as seizures from temporal or frontal nocturnal epilepsy, with which they share local patterns of brain activation at night.

Sleep walking is characterized by complete or abortive walking behavior occurring abruptly at night out of SWS, prompted by an incomplete spontaneous arousal. At the end of the episode the patient spontaneously resumes nocturnal sleep in bed or elsewhere and does not recall his behavior in the morning or even immediately post hoc when awakened.

Functional neuroimaging (SPECT) of the sleep walking episode demonstrates activation of thalamo-cingulate pathways, paralleled by decreased regional cerebral blood flow in the fronto-parietal cortices, as if the brain were motorically activated but unconscious due to the switch-off of the executive functions [39].

Exactly the same areas are at stake during paroxysmal arousals of nocturnal frontal lobe epilepsy (NFLE) [40]. Although the areas of activation are the same, the patterns of EEG activity differ as demonstrated by intracerebral recordings of presurgically monitored epileptic patients [41-42]. The motor and cingulate cortex contact electrodes show activation whereas persistent slow activity is observed over the fronto-parietal associative cortices. The uncoupling of EEG rhythms represent one more example of local partial sleep and indicate a different threshold of arousability between the motor-cingulate and the associative cortices, which appear to struggle to maintain the continuity of sleep while other brain areas are moving towards arousal. Thus an enhanced delta power over the frontal regions of the brain in the presence of a local motor arousal would explain the specific lack of awareness and insight. Similar EEG/brain activity dissociation supports confusional arousals both in children and, more rarely, in adults in the form of sleep inertia or sleep drunkenness [43].

Basic drives may complicate these episodes with the occurrence of sleep-related eating disorder (SRED) or complex auto or hetero sex behaviors (sexsomnia) [44], all of which is performed by the subject while still partially asleep and therefore in an altered state of consciousness which is not morally and legally prosecutable [45]. During night terrors there is, instead, an intense autonomic activation including midryasis, diaphoresis, tachycardia and tachypnea. Screaming, vocalization or inconsolable crying usually alert parents or bed partners. Any attempt to stop or console the patient is fruitless and may actually only prolong the episode, which as a rule, will be forgotten in the morning by the unaware child. Night terrors are very rare in adults (2,7%) and usually represent reemergence of childhood behaviors, whereas sleepwalking has a higher life time (22,4%) but lower adult prevalence (1,7%) [46].

REM Parasomnias

Parasomnias occurring out of REM sleep represent further examples of sleep-wake dissociation. This is true in particular for hypnagogic hallucinations and paralysis (see narcolepsy above) which may also occur sporadically in asymptomatic young subjects.

REM behavior disorder (RBD) is an interesting reverse example of motor activation during REM sleep taking place instead of the physiological state trait of motor inhibition. Therefore dreams, especially violent dreams, may become "enacted" due to the impairment of the physiologic REM related inhibition of muscle tone, thereby causing self or hetero injurious behaviors during sleep.

This disorder was first described in 1986 by Schenck and Mahowald [47] in elderly males and later characterized as acute or chronic, idiopathic or cryptogenic, although more recently seen as a precursor of neurodegenerative central nervous system (CNS) disorders, mainly synucleinopathies [48]. Acute forms of the disease may be iatrogenic due to intoxication or withdrawal from substances/medications such as alcohol, barbiturates or antidepressants.

Alteration of the brainstem motor networks responsible for motor inhibition (SLD, perilocus coeruleus and/or long descending fibers from the forebrain), represent the anatomical basis of RBD. PET and SPECT studies of RBD patients have documented impaired dopamine transporter (DAT) function in the nigro-striatal system of the brainstem akin, but with lesser extent, to abnormalities seen in Parkinson Disease patients.

Video-polygraphic recordings allow instrumental confirmation of the clinical diagnosis via detection of REM sleep without atonia.

Subtypes and variants of RBD include:

Parasomnia overlap disorder (POD) consisting of a combination in the same subjects of DOA with RBD, predominantly in younger males, idiopathic or secondary to known CNS disease from brain tumors to multiple sclerosis (MS) and psychiatric disorders.

The extreme form of RBD+ condition is status dissociatus (SD), a clinical condition of extreme continuous state dissociation where dream-like and delirious agitated behaviors occur during apparent sleep, albeit no conventional sleep stages may be classified due to complete loss of any kind of physiologic state markers [49].

Extreme sympathetic over- activation is often responsible for premature death, especially in HIV brainstem involvement or in Fatal Familial Insomnia (FFI), a prion disease which may manifest as SD during its progressive course.

Sleep itself and sleep disorders may indeed teach us a lot about altered states of consciousness and may be useful to better understand and model conditions typical of psychiatric and comatose patients.

References

[1] Zeman A. Consciousness: concepts, neurobiology, terminology of impairments,theoretical models and philosophical background. *Handb. Clin. Neurol.* 2008;90:3-31.

[2] Zeman A. What do we mean by "conscious" and "aware"? *Neuropsychol. Rehabil.* 2006;16:356-76.

[3] Laureys S, Perrin F, Brédart S. Self-consciousness in non-communicative patients. *Conscious. Cogn.* 2007;16:722-41;

[4] National Clinical Guideline Centre for Acute and Chronic Conditions (UK). Transient Loss of Consciousness ('Blackouts') Management in Adults and Young People [Internet]. London: Royal College of Physicians (UK); 2010.

[5] van Dijk N, Sprangers MA, Colman N, Boer KR, Wieling W, Linzer M. Clinical factors associated with quality of life in patients with transient loss of consciousness. *J. Cardiovasc. Electrophysiol.* 2006;17:998-1003.

[6] Blumenfeld H. Impaired consciousness in epilepsy. *Lancet Neurol.* 2012;11:814-26.

[7] Berg AT, Cross JH. Classification of epilepsies and seizures: historical perspective and future directions. Handb Clin Neurol. 2012;107:99-111.

[8] McLauchlan DJ, Robertson NP. Management of status epilepticus. *J. Neurol.* 2012;259:2261-3.

[9] Margineanu DG. Epileptic hypersynchrony revisited. *Neuroreport.* 2010;27;21:963-7.

[10] Gallentine WB, Mikati MA. Genetic generalized epilepsies. *J. Clin. Neurophysiol.*2012; 29:408-19.

[11] Hogan RE, English EA. Epilepsy and brain function: common ideas of Hughlings-Jackson and Wilder Penfield. *Epilepsy Behav.* 2012;24:311-3.

[12] Penfield W. The physiology of epilepsy. *Adv. Neurol.* 1975;8:1-9.

[13] Sahaya K, Dholakia SA, Sahota PK. Psychogenic non-epileptic seizures: a challenging entity. *J. Clin. Neurosci.* 2011;18(12):1602-7.

[14] Attard A, Ranjith G, Taylor D. Delirium and its treatment. *CNS Drugs.*2008;22:631-44.

[15] Heskestad B, Baardsen R, Helseth E, Ingebrigtsen T. Guideline compliance in management of minimal, mild, and moderate head injury: high frequency of non-compliance among individual physicians despite strong guideline support from clinical leaders. *J. Trauma.* 2008;65:1309-13.

[16] Zandbergen EG. Postanoxic coma: how (long) should we treat? *Eur. J. Anaesthesiol. Suppl.* 2008;42:39-42.

[17] Zandbergen EG, Koelman JH, de Haan RJ, Hijdra A; PROPAC-Study Group. SSEPs and prognosis in postanoxic coma: only short or also long latency responses? *Neurology.* 2006 22;67:583-6.

[18] Wijdicks EF. The neurologist and Harvard criteria for brain death. *Neurology.* 2003;61:970-6.

[19] Scripko PD, Greer DM. An update on brain death criteria: a simple algorithm with complex questions. *Neurologist.* 2011;17:237-40.

[20] Webb A, Samuels O. Brain death dilemmas and the use of ancillary testing. *Continuum* (Minneap Minn). 2012;18:659-68.

[21] Nakagawa TA, Ashwal S, Mathur M, Mysore M; Committee For Determination Of Brain Death In Infants Children. Guidelines for the determination of brain death in infants and children: an update of the 1987 task force recommendations-executive summary. *Ann. Neurol.* 2012;71:573-85.

[22] Bernat JL. Chronic disorders of consciousness. *Lancet.* 2006;367:1181-92.

[23] Formisano R, D'Ippolito M, Risetti M, Riccio A, Caravasso CF, Catani S, Rizza F, Forcina A, Buzzi MG. Vegetative state, minimally conscious state, akinetic mutism and Parkinsonism as a continuum of recovery

from disorders of consciousness: an exploratory and preliminary study. *Funct. Neurol.* 2011;26:15-24.

[24] Hirschberg R, Giacino JT. The vegetative and minimally conscious states: diagnosis, prognosis and treatment. *Neurol. Clin.* 2011;29:773-86.

[25] Sarà M, Pistoia F. Bedside detection of awareness in the vegetative state. *Lancet.* 2012;379:1702-3.

[26] Borthwick CJ, Crossley R. Permanent vegetative state: usefulness and limits of a prognostic definition. *NeuroRehabilitation.* 2004;19:381-9.

[27] Giacino JT, Ashwal S, Childs N, Cranford R, Jennett B, Katz DI, Kelly JP, Rosenberg JH, Whyte J, Zafonte RD, Zasler ND. The minimally conscious state: definition and diagnostic criteria. *Neurology.* 2002;58:349-53.

[28] Kotchoubey B, Merz S, Lang S, Markl A, Müller F, Yu T, Schwarzbauer C. Global functional connectivity reveals highly significant differences between the vegetative and the minimally conscious state. *J. Neurol.* 2012. [Epub ahead of print].

[29] von Wild K, Laureys ST, Gerstenbrand F, Dolce G, Onose G. The vegetative state--a syndrome in search of a name. *J. Med. Life.* 2012;5:3-15.

[30] Laureys S, Celesia GG, Cohadon F, Lavrijsen J, León-Carrión J, Sannita WG, Sazbon L, Schmutzhard E, von Wild KR, Zeman A, Dolce G; European Task Force on Disorders of Consciousness. Unresponsive wakefulness syndrome: A new name for the vegetative state or apallic syndrome. *BMC Med.* 2010;8:68.

[31] Shetty AC, Morris J, O'Mahony P. Akinetic mutism--not coma. *Age Ageing.* 2009;38:350-1.

[32] Lulé D, Zickler C, Häcker S, Bruno MA, Demertzi A, Pellas F, Laureys S, Kübler A. Life can be worth living in locked-in syndrome. *Prog. Brain Res.* 2009;177:339-51.

[33] Werner P, Karnieli-Miller O, Eidelman S. Current knowledge and future directions about the disclosure of dementia: A systematic review of the first decade of the 21st century. *Alzheimers Dement.* 2012. pii:S1552-5260(12)00082-9.

[34] Hobson, J. A. REM sleep and dreaming: towards a theory of protoconsciousness. *Nature Reviews Neuroscience* 2009 Nov; 10: 803-13.

[35] Vyazovskiy, V. V., Olcese, U., Hanlon, E. C, Nir, Y., Cirelli, C., Tononi, G. Local sleep in awake rats. *Nature* 2011 Apr; 472: 443-447.

[36] Nobili, L., De Gennaro, L., Proserpio, P., Moroni, F., Sarasso, S., Pigorini, A., De Carli, F., Ferrara, M. Local aspects of sleep: observations from intracerebral recordings in humans. *Prog. Brain Res.* 2012;199: 219-32.

[37] Mahowald, M. W., Schenck, C. H. Insights from studying human sleep disorders. *Nature* 2005; 437: 1279-85.

[38] Mayer, G. (2012) Narcolepsy. In Silvestri R (Ed) Sleep Disorders in Neurology. Nova Science Pub. Inc. New York, pp. 17-28.

[39] Bassetti, C., Vella, S., Donati, F., Wiellep, P., Weder, B. SPECT during sleep walking. *Lancet* 2000; 356: 484-5.

[40] Vetrugno, R., Mascalchi, M., Vella, A., et al. Paroxysmal arousal in epilepsy associated with cingulate hyperperfusion. *Nuerolog*y 2005; 64: 356-358.

[41] Terzaghi, M., Sartori, I., Tassi, L., Didato, G., Rustioni, V., Lo Russo, G., Manni, R., Nobili, L. Evidence of dissociated arousal states during NREM parasomnia from an intracerebral neurophysiological study. *Sleep* 2009; 32: 409-412.

[42] Terzaghi M, Sartori I, Tassi L, Rustioni V, Proserpio P, Lorusso G, Manni R, Nobili L. Dissociated local arousal states underlying essential clinical features of non-rapid eye movement arousal parasomnia: an intracerebral stereo-electroencephalographic study. *J. Sleep Res.* 2012 Oct; 21(5):502-6.

[43] Stores, G. (2010) Confusional arousals. In Thorpy MJ, Plazzi G (Eds) *The parasomnias and other sleep-related movement disorders.* Cambridge University Press. New York, pp. 99-108.

[44] Schenck, C. H., Arnulf, I., Mahowald, M. W. Sleep and sex: what can go wrong? A review of the literature on sleep related disorders and abnormal sexual behaviors and experiences. *Sleep* 2007; 30: 683-702.

[45] Mahowald, M. W., Schenck, C. H. Parasomnias: sleep walking and the law. *Sleep Med. Review* 2000; 4: 321-39.

[46] Bjorvatn, B., Gronli, J., Pallesen, S. (2010) Prevalence of different parasomnias in the general population. *Sleep Med.* 11: 1031-1034.

[47] Schenck, C. H., Bundlie, S., Ettinger, M., Mahowald, M. W. Chronic behavioral disorders of human REM sleep: a new category of parasomnia. *Sleep* 1986; 9: 293-308.

[48] Boeve, B., Silber, M., Parisi, J., et al. Synucleinopathy pathology and REM sleep behavior disorder plus dementia or parkinsonism. *Neurology* 2003; 61: 40-5.

[49] Mahowald, M. W., Schenck, C. H. Status dissociatus a perspective on states of being. *Sleep* 1991; 14: 69-79.

In: Chronic Disorders of Consciousness ISBN: 978-1-62618-940-9
Editors: R. S. Calabrò and P. Bramanti © 2013 Nova Science Publishers, Inc.

Chapter 3

Clinical and Instrumental Evaluation of Chronic Disorders of Consciousness

Serena Gennaro[1], Nunzio Muscarà[2], Antonio Leo[2], Simone Reitano[2], David Militi, Placido Bramanti[2] and Rocco Salvatore Calabrò[2]
[1]Centro Cardinal Ferrari, Fontanellato, Italy
[2]IRCCS Centro Neurolesi "Bonino-Pulejo", Messina, Italy

Abstract

With the increasing level of intensive care in modern time available, more people survive to brain injury, developing a chronic disorder of consciousness such as vegetative state (SV) and minimally conscious state (MCS). The most important difference between the two conditions is the presence or the absence of definite behavioral evidence of self or environmental awareness. Several studies have been performed to identify reliable tools to assess the presence of awareness and to allow differential diagnosis between VS and MCS. The most important clinical evaluation tool is represented by the Coma Recovery Scale-Revised (CRS-R) composed of six sub-scale addressing auditory, visual, motor, oromotor/verbal, communication and arousal processes. Since the bedside examination of consciousness in severely brain damaged patients is very challenging and often complicated by the presence of underlying deficits in the domain of verbal or non-verbal communication functions, paraclinical measures have been also identified. EEG standard, sleep EEG and EPs present characteristic features that distinguish VS from MCS. In the last period, fRMI studies have offered the possibility to measure directly and non-invasively severely brain damaged patients' brain activity and show that at least some patients, who appear unresponsive, may retain higher levels of self and environmental awareness than would otherwise be suggested on bedside examination.

Introduction

As our understanding of the complex mechanisms of brain injury continues to develop, accompanied by advances in medical science and technology, the numbers of patients surviving severe brain injury is increasing.

Whilst some who make it to critical care units emerge from coma, which is often the first response to brain injury, a significant number remain unresponsive, showing a chronic disorder of consciousness.

The best-known definition of coma is that proposed by Jennett and Teasdale [1], which describe the patient as having ' no verbal response, not obeying commands and not opening the eyes spontaneously or to stimulation'. The literature on coma also describes patients as having no arousal and no awareness, no eyes opening, reflex movements only, no language comprehension and no 'purposeful' response. Coma is a short-term condition and usually in 2-4 weeks most patients either regain consciousness, die or evolve towards a chronic disorders of consciousness, represented by vegetative state (VS) and minimally conscious state (MCS) [2]. These states are characterized by absence of behavioural responses that is considered to equate to lack of consciousness.

Wakefulness Without Awareness

Consciousness is an ambiguous term, encompassing both wakefulness and awareness. The presence or the absence of consciousness is fundamental to distinguish VS from MSC.

The term wakefulness refers to a state in which the eyes are open and there is a degree of motor arousal; it contrasts with sleep, a state of eyes closure and motor quiescence. Different degrees could be recognized and wakefulness is normally associated with conscious awareness. On the other hand, the same existence of VS indicates that wakefulness and awareness can be dissociated. This can occur because the brain systems that control wakefulness, located in the upper brainstem and thalamus, are largely distinct from those that mediate awareness [3].

With the term awareness we indicate the ability to have experiences of any kind. We are typically aware of our surroundings and of bodily sensations, but the contents of awareness can also include our memories, thoughts, emotions and intentions. Awareness is not a single indivisible capacity and brain damage can selectively impair some aspects of awareness, leaving others intact [4].

There is no simple single clinical sign or laboratory test of awareness. Its presence must be deduced from a range of behaviours, which indicate that an individual can perceive self and surroundings, frame intentions and communicate.

Chronic Disorders of Consciousness

The Vegetative State

A patient in the vegetative state (VS) appears at times to be wakeful, with cycles of eyes closure and opening, resembling those of sleep and waking. However, close observation reveals no sign of awareness or of a 'functioning mind'. Specifically there is no evidence that the patient can perceive the environment or his/her own body or communicate with others. As a rule, the patient can breathe spontaneously and has a stable circulation. VS is often described as 'loss of function of the cortex while the function of the brainstem and allied structures is preserved' [5].The functional preservation of these structures maintains arousal and autonomic functions in these patients [6].

Jennet and Plum chose the term "vegetative state" as a description of the patient's behaviour, to indicate 'a body capable of growth and development but devoid of sensation and thought' [7].

A vegetative state persisting for a month following TBI or non-traumatic brain damage is referred to as a "persistent vegetative state", while a vegetative state that persists for 3 months following a non-traumatic brain injury (such as hypoxic-ischemic, metabolic and congenital) or 12 months following TBI is regarded as irreversible and thus termed a "permanent vegetative state" [8].

Patients in the VS may be aroused by painful or salient stimuli, but they demonstrate no clear signs of conscious perception or deliberate action [3].

Stereotyped responses to external stimuli, such as grimacing, are thought to result from preserved brainstem circuits and limbic cortical regions [6].

At least patients in VS show no signs of being aware of themselves or their environment and have no evidence of language comprehension or expression using verbal or non-verbal signals [9].

Minimally Conscious State

The MCS is a condition of severely altered consciousness in which minimal but definite behavioral evidence of self or environmental awareness is demonstrated.

MCS is characterized by behaviors associated with conscious awareness that occur inconsistently, but are reproducible or sustained long enough to discern their nature. Patients in a MCS appear similar to VS patients, exhibiting an inability to communicate consistently but, in contrast to VS, they show limited yet clear evidence of awareness, on a reproducible or sustained basis.

To summarize, VS and MCS patients share the following in commons:

a. sufficiently preserved hypothalamic and brainstem autonomic functions to permit prolonged survival with medical and nursing care;
b. preserved respiratory functions;
c. intermittent wakefulness manifested by the preservation of sleep-wake cycles;
d. bowel and bladder incontinence;
e. variably preserved cranial nerve and spinal reflexes;
f. severe impairment of motor function.

However, they also present several important differences.

Whereas patients in VS have no awareness of self or the environment, MCS patients have definite, although limited, awareness of self or the environment and shown purposeful behaviour, including reactions to environmental stimuli that are not reflex actions [10].

MCS patients may show a limited but reproducible form of simple communication and are also more likely to experience pain and suffering, in contrast to VS patients who by definition are unable to do so.

It is important to underlying that persons suspected to be MCS have to be carefully examined to be distinguished from those with locked-in syndrome. These because both conditions are characterized by minimal behavioral interaction with the environment and they can be easily misdiagnosed [11].

Clinical Evaluation

The distinction between the VS and the MCS may be subtle and requires careful evaluation [12]. Beginning with clinical signs, there are structured criteria for diagnosis of VS and MCS.

In the evaluation of suspected VS, first of all has to be excluded the possibility of effects of sedative, anaesthetic or neuromuscular blocking drugs and the possibility that continuing metabolic disturbance is responsible for the clinical features. Then has to be met the following criteria, which represent the key requirements for the diagnosis:

- there must be no evidence of awareness of self or of environment at any time; no response to visual, auditory, tactile or noxious stimuli of a kind suggesting volition or conscious purpose; no evidence of language comprehension or meaningful expression;
- can be shown the typically cycles of eye closure and opening, giving the appearance of sleep-wake cycle;
- can be shown evidences of sufficiently preserved hypothalamic and brainstem function, to ensure the maintenance of respiration and circulation.

Patients in VS usually satisfy the second and the third signs but, unlike the first criterion, they are not obligatory [13].

For example, a patient with cerebral injuries, sufficient to cause the VS, might, incidentally, have the III nerve palsies preventing eye opening, or injuries to the chest or medulla affecting respiratory function. In these cases the second and the third signs may not be respected.

Beside signs of a cycle of sleep and wakefulness, patients in VS may make a range of spontaneous movements including chewing, teeth grinding, swallowing, roving eye and purposeless limb movements or facial movements such as grimaces, shed tears, or make grunting or groaning sounds for no discernible reason [14].

Table 1. Comparison of clinical features associated with vegetative state and minimally conscious state

Condition	consciousness	Sleep/ wake	Motor function	Auditory function	Visual function	Communication	Emotion
Vegetative State	None	present	Postures or withdraws to noxious stimuli; occasional non purposeful movement	Startle; Brief orienting to sound	Startle; Brief visual fixation	None	None; Reflexive crying or smiling
Minimally conscious state	None	present	Localizes noxious stimuli; reaches for objects; holds or touches objects in a manner that accommodates size and shape; automatic movements	Localizes sound location; inconsistent command following	Sustained visual fixation; sustained visual pursuit	Contingent vocalization; Inconsistent but intelligible verbalization or gesture	Contingent smiling and crying

There could be reactions to a number of stimuli: brainstem reflexes (papillary, oculocephalic, corneal, oculovestibular) or generalized arousal response and extensor or flexor withdrawal of a limb caused by noxious and noisy stimuli.

On the other side, the distinction of MCS from VS is strictly related to the presence of behaviours associated with conscious awareness. In MCS, cognitively mediated behavior occurs inconsistently, but is reproducible or sustained long enough to be differentiated from reflexive behavior.

The reproducibility of such evidence is affected by the consistency and complexity of the behavioral response. Extended assessment may be required to determine whether a simple response, (for example finger movement or eye blink) that is observed infrequently, is occurring in response to a specific environmental event or on a coincidental basis. In contrast, a few observations of a complex response may be sufficient to determine the presence of consciousness [15].

To make the diagnosis of MCS, limited but clearly discernible evidence of self or environmental awareness must be demonstrated on a reproducible or sustained basis by one or more of the following behaviors:

- following simple commands,
- gestural or verbal yes/no responses (regardless of accuracy),
- intelligible verbalization,
- purposeful behavior, including movements or affective behaviors that occur in contingent relation to relevant environmental stimuli and are not due to reflexive activity. Some examples of qualifying purposeful behavior include: appropriate smiling or crying in response to the linguistic or visual content of emotional, but not to neutral topics or stimuli; vocalizations or gestures that occur in the direct response to the linguistic content of questions; reaching for objects that demonstrates a clear relationship between object location and direction of reach; touching or holding objects in a manner that accommodates the size and the shape of the object; pursuit eye movement or sustained fixation that occurs in direct response to moving or salient stimuli [16].

To better avoid equivocal behavioral responses of patients suspected of being in MCS, has to be followed specific recommendations to facilitate detection of conscious awareness [17]:

- adequate stimulation should be administered to assure that arousal in maximized;
- factors adversely affecting arousal should be addressed (sedating medications, seizures);
- attempts to elicit behavioral responses through verbal instruction should not involve behaviors that frequently occur on a reflexive basis;
- command-following trials should incorporate motor behaviors expected to be within the person's capability;
- a variety of different behavioral responses should be investigated using a broad range of eliciting stimuli
- examination procedures should be conducted in a distraction-free environment;
- serial reassessment incorporating systematic observation and measurement strategies should be used to confirm the validity of the initial assessment. Specialized tools and procedures designed for quantitative assessment may be useful (table 1).

Coma Recovery Scale – Revised (CRS-R)

The CRS-R represents the best-standardized tool for the clinical assessment of chronic disorders of consciousness, mostly to distinguish VS from MSC [18].

The scale was first described in 1991 and revised in 2004 [19, 20]. It consists of six sub-scale addressing auditory, visual, motor, oromotor/verbal, communication and arousal processes.

The lowest score on each sub-scale represents reflexive activity, while the highest score describes cognitively mediated behaviours. The total score ranges from 0, the worst, to 23, the best (table 2).

Recently, according to the different scores obtained, MCS was subcategorized: MCS- describes low-level behavioural responses (visual pursuit, localization of noxious stimulation or contingent behaviour such as appropriate smiling or crying to emotional stimuli) and MCS+ describes high-level behavioural responses (command following, intelligible verbalizations or non-functional communication) [21].

The scale expressly incorporates diagnostic criteria for VS, MCS and the emergence from MCS (MCS+), allowing the examiner to make a diagnosis based directly on bedside behavioural observations (table 2b).

Table 2

Coma Recovery Scale-Revised
Auditory function
4 - Consistent Movement to Command*
3 - Reproducible Movement to Command*
2 - Localizaion to Sound
1 - Auditory Startle
0 - None
Visual function
5 - Object Recognition*
4 - Object Localization: Reaching*
3 - Visual Pursuit *
2 - Fixation *
1 - Visual Startle
0 - None
Motor function
6 - Functional Object Use+
5 - Automatic Motor Response*
4 - Object Manipulation*
3 - Localizaion to Noxious Stimulation*
2 - Flexion Withdrawal
1 - Abnormal Posturing
0 - None/Flaccid
Oromotor/Verbal function
3 - Intelligible Verbalization*
2 - Vocalization/Oral Movement
1 - Oral Reflexive Movement
0 - None
Communication
2 - Functional: Accurate+
1 - Non-Functional: Intentional*
0 - None
Arousal
3 - Attention
2 - Eye Opening without Stimulation
1 - Eye Opening with Stimulation
0 - Unarousable

+Denotes emergence from minimally conscious state.
*Denotes minimally conscious state.

Table 2b. Scores in vegetative state (VS), minimally conscious state (MCS) and emergence from the minimally conscious state (MCS+) according to the sub-scales of the Coma Recovery Scale-Revised

Sub-scales	VS	MCS	MCS+
Auditory	< o = 2	3-4	3-4
Visual	< o = 1	2-5	2-5
Motor	< o = 2	3-5	6
Oromotor/verbal	< o = 2	3	3
Communication	0	1	2

If possible, it is important to exclude aphasia, agnosia, apraxia or sensorimotor impairment as the basis for non responsiveness.

The scale has shown to have adequate reliability and validity [22] and has been used to investigate, not only clinical diagnostic accuracy [23], but also outcome prediction [24], treatment effectiveness and relationship between behavioral and neurophysiologic markers of consciousness [25, 26, 27].

Instrumental Evaluation

The bedside examination of consciousness in severely brain damaged patients often is very challenging because responses may be very small, inconsistent and easily exhausted.

This issue is further complicated when patients have underlying deficits in the domain of verbal or non-verbal communication functions, such as aphasia, agnosia or apraxia [28].

So the detection of volitional behavior in patients with severe brain injury is often difficult, as fluctuations in arousal level, deficient drive and sensory and motor impairments can mask signs of consciousness.

Thus an objective evaluation providing a paraclinical measure could help the differential diagnosis between VS and MCS.

EEG

Cortical activity and electrophysiological patterns are classically related mainly to brainstem function (through the ascending reticular activating system) and thalamocortical function [29]. So detectable scalp EEG changes reflect the activity of ascending reticular system, and, also, may investigate the integrity and the function of thalamocortical connections [30].

Different studies conducted have shown differences in EEG resting-activity between VS and MSC [31].

Substantially, the quantitative analysis of EEG shows a higher presence of delta activity in VS than in MSC. However, a drop in alpha presence was observed in VS compared to MCS. Also connectivity measurements showed that MCS patients had a better-connected network in the theta and in the alpha bands.

No differences were found in the traumatic vs non-traumatic setting [32].

Sleep EEG

It is known that the presence of standard sleep elements reflects global brain integrity, as they have been shown to be altered in several pathological states [33, 34].

Also it has been identified a relationship between the presence of normal sleep electrophysiological features and the level of consciousness in severely brain damaged patients [35]. Thus, the presence of normal sleep patterns, likely reflecting underlying functional brain integrity, could differentiate the VS patients from those in MCS.

A study has shown that VS patients do not present any electrophysiological features of normal sleep, even if preserved behavioural sleep-wake cycles could be observed [36].

Even if transient periods of behavioural wakefulness (increased arousal, eyes opening) were observed, no related EEG change neither REM sleep patterns could be detected in those patients. In contrast, subjects in a minimally conscious state show all stages of sleep.

The presence of spindles also suggests that patients in minimally conscious state show a relative preservation of thalamocortical connectivity.

Moreover it has been observed that MCS patients have detectable sleep cycle architecture and homoeostatic regulation of slow-wave sleep.

EPs

The evoked potentials (EPs) are used in the diagnosis but mostly in the prognostic evaluation of chronic disorders of consciousness and they result mostly altered in VS patients. Although brainstem auditory evoked potentials (BAEPs) can be normal in VS, BAEPs waveform is often attenuated, delayed or absent, depending on the location and degree of brainstem injury.

Also somatosensory evoked potentials (SEPs) have been found to be abnormal in VS patients with delay, attenuation, or even absence of the N20 cortical response to median nerve stimulation [37].

fRMI

In the past 15 years neuroimaging studies have offered the possibility to measure directly and non-invasively severely brain damaged patients' brain activity. More recently these techniques have been developed aiming to detect neural motor independent command following. A study about fMRI, conducted by the Cambridge group, showed how a patient, considered to be in VS, was able to activate the supplementary motor area following the request to imagine playing tennis. Instead the instruction to imagine moving around in her house resulted in parahippocampal activation [38].

Since this case report similar active or command following paradigms, severely brain damaged patients have been tested using fMRI. In a next step, Monti and Vanhaudenhuyse [39] used the technique in a large group of patients with disorders of consciousness and established a fRMI-based communication. The patients were asked to answer autobiographical questions by doing motor imagery (such as playing tennis) to answer 'yes' and visuo-spatial imagery to answer 'no' (such as walking around the rooms of their house). Out of 54 patients enrolled in the study, five patients demonstrated willful modulation of brain activity (activating supplementary motor or parahippocampal areas, depending on the command). Four of these patients were admitted to the hospital with a diagnosis of VS. However, when assessed with the CRS-R, some behavioural indicators of consciousness could be detected in two cases.

In others recent studies have been shown how was possible to activate, in patient with diagnosis of VS, a specific brain area: activation of language brain network at the request of silently naming the object appearing in the screen [21], activation of premotor cortex at the request to move the hand [40] or activation of supplementary motor area imagining to swim [41, 42].

These referenced studies indicate that at least some patients, who appear unresponsive, may retain higher levels of self and environmental awareness than would otherwise be suggested on bedside examination. As such, the burden of proof for establish consciousness in severely brain damaged patients no longer exclusively lies in behavioural assessment and functional neuroimaging paradigms may eventually play an important role in reducing diagnostic error and improve prognostic accuracy.

For these reason Owen and colleagues has proposed to identified a new diagnostic category, since patients whose use a complex mental imagery task to accurately communicate cannot be considered vegetative or minimally conscious.

They suggested to call this condition " functional locked-in syndrome" to emphasize the dissociation between their extreme behavioural motor dysfunction and the preserved cognitive functions as shown by functional imaging techniques [43].

Conclusion

These evidences have opened a discussion about the importance of fRMI evaluation and the limitation of behavioural assessment. Otherwise, in some case, the CRS-R was able to distinguish signs of consciousness in patients clinically evaluated as VS.

So at the moment the difficulties to use fRMI, in consideration of its not largely diffusion, impose the use of clinical and other para-clinical instrument (such as CRS-R, EEG and EPs) in the assessment of chronic disorder of consciousness.

References

[1] Jennett B, Teasdale G., Aspects of coma after severe head injury. *Lancet,* (1977); 1: 878-81

[2] Shiel A, Gelling L., Wilson B, Coleman M, Pickard JD Difficulties in diagnosis the vegetative state. *British Journal of Neurosurgery,* (Febbraio 2004); 18(1): 5-7.

[3] Zeman A. Consciousness *Brain*, (2001); 124: 1263-89.

[4] Laureys S, Berre J, Goldman S Cerebral function in coma, vegetative state, minimally conscious state, locked-in syndrome and brain death. *Yearbook of intensive care and emergency medicine,* 2001: 386-96.

[5] Adams JH, Graham DI, Jennett B. The neuropathology of the vegetative state after an acute brain insult. *Brain*, (2000)123; 1327-1338.

[6] Maquet P, Degueldre C, Delfiore G, Aerts J, Peters JM, Luxen A et al Functional neuroanatomy of human slow wave sleep, *Journal of Neuroscience,* 17 (1997), 2087-2812.

[7] Jennett B and Plum F Persistent vegetative state after brain damage, a syndrome in search of a name. *Lancet,* (1972) 1; 734-737.

[8] The Multi society task force of persistent vegetative state, Medical aspects of the persistent vegetative state. *New England Journal of Medicine,* (1994) 330; 1572-79.

[9] Owen AM, Coleman MR, Using neuroimaging to detect awareness in disorders of consciousness, *Functional Neurology,* (2008) 23, 189-194.

[10] Giacino JT, Ashwal S, Childs N, Cranford R, Jennett B, Katz DI, The minimally conscious state: definition and diagnostic criteria, *Neurology,* (2002) 58, 349-53.

[11] Giacino JT, Kalmar K The vegetative and minimally conscious states: a comparison of clinical features and functional outcame. *J. head trauma rehabilitation,* (1997) 12: 36-51.

[12] Schnakers C, Vanhaudenhuyse A, Giacino J, Ventura M, Boly M, Majerus S, Moonen G, Laureys S. Diagnostic accuracy of the vegetative and minimally conscious state: clinical consensus versus standardized neurobehavioral assessment. *BMC Neurology,* (2009); 9-35.

[13] Jennett B, Plum F. Persistent vegetative state after brain damage. *Lancet,* (1972); i:734-7.

[14] Schiff ND, Ribary U, Moreno DR, Beattie B Residual cerebral activity and behavioural fragments can remain in the persistently vegetative brain. *Brain*, (2002); 125: 1210-34.

[15] Childs NL, Mercer WN, Childs HW. Accurancy of diagnosis of persistent vegetative state. *Neurology,* (1993); 43: 1465-67.

[16] Whyte J, DiPasquale MC. Assessment of vision and visual attention in minimally responsive brain injury patients. *Arch. Phys. Med. Rehabil.,* (1995); 765: 804-10.

[17] The Aspen Neurobehavioral Conference Work Group. The minimally conscious state: diagnosis and prognosis. *Neurology,* (2002).

[18] Teasdale G, Jennett B. Assessment of coma and impaired consciousness. A practical scale. *Lancet,* (1974); 2:81-84.

[19] Giacino JT, Kezmarsky MA, DeLuca J, Cicerone KD. Monitoring rate of recovery to predict outcome in minimally responsive patients. *Arch. Phys. Med. Rehabil.,* (1991); 72:897-901.

[20] Giacino J, Kalmar K, Whyte J. The JFK coma recovery scale-revised: measurement characteristics and diagnostic utility. *Archives of Physical Medicine and Rehabilitation,* (2004); 85: 2020-29.

[21] Rodriguez Moreno D, Schiff ND, Giacino J, Kalmar K, Hirsch J. A network approach to assessing cognition in disorders of consciousness. *Neurology*, (2010) 75: 1871-78.

[22] Lombardi F, Gatta G, Sacco S, Muratori A, Carolei A. The Italian version of the Coma Recovery Scale-Revised (CRS-R).*Functional Neurology,* (2007); 22:47-61.

[23] Schnakers C, Majerus S, Giacino J. A French validation study of the coma recovery scale-revised. *Brain Inj.,* (2008); 22(10):786-92.

[24] Vanhaudenhuyes A, Giacino J, Schnakers C. Blink to visual threat does not herald consciousness in the vegetative state. *J. Neurol. Neurosurg. Psychiatry,* (2008); 79(2):225-7.

[25] Coleman MR, Rodd JM, Davis MH. Do vegetative patients retain aspects of language comprehension? Evidence from fRMI. *Brain,* (2007);130:2494-507.

[26] Vanhaudenhuyse A, Schnakers C, Bredart S. Assessment of visual pursuit in post-comatose states: use a mirror. *J. Neurol. Neurosurg. Psychiatry,* (2008); 79(2):223.

[27] Smart C, Giacino J, Cullen T. Locked-in syndrome complicated by central deafness: neuropsychological and neuroimaging findings. *Nat. Clin. Pract. Neurol.,* (2008); 4(8): 448-53.

[28] Majerus S, Bruno MA, Schnakers C, Giacino JT, Laureys S. The problem of aphasia in the assessment of consciousness in brain-damage patients. *Prog. Brain Res.,* (2009) 177:49-61.

[29] Boly M, Phillips C, Tshibanda L, Vanhaudenhuyse A, Schabus M, Dangvu TT. Intrinsic brain activity in altered states of consciousness: how conscious is the default mode of brain function? *Ann. NY Acad. Sci.,* (2008); 1129: 119-29.

[30] Vanhaudenhuyse A, Noirhomme Q, Tshibanda J, bruno MA, Boveroux P, Schnakers C. default network connectivity reflects the level of consciousness in non-communicative brain damaged patients. *Brain* (2010); 133 (Pt 1): 161-71.

[31] Leon-Carrion J, Martin-Rodriguez JF, Damas-Lopez J, Barroso Martin JM, Dominguez-Morales MR. Brain function in the minimally conscious state: a quantitative neurophysiological study. *Clin. Neurophysiol.,* (2008); 119:1509-14.

[32] Lehembre R, Bruno MA, Vanhaudenhuyse A, Chatelle C, Cologan V et all. Resting-state EEG study of comatose patients: a connectivity and frequency analysis to find differences between vegetative and minimally conscious states. *Funct. Neurol.,* (2012); 27(1): 41-47.

[33] Bassetti CL, Aldrich MS. Sleep electroencephalogram changes in acute hemispheric stroke. *Sleep Med.,* (2001); 2:185-94.

[34] Crowley K, Sullivan EV, Adalsteinsson E, Pfefferbaum A, Colrain IM. Differentiating pathologic delta from healthy physiologic delta in patients with Alzheimer disease. *Sleep Med.,* (2005); 28:865-70.

[35] Cologan V, Schabus M, Ledoux D, Moonen G, Maquet P, Laureys S. Sleep in disorders of consciousness. *Sleep Med. Rev.,* (2010); 14:97-105.

[36] Landsness E, Bruno MA, Noirhomme Q, Riedner B, Gosseries O, Schnakers C, Massimini M et all. Electrophysiological correlates of behavioural changes in vigilance in vegetative state and minimally conscious state. *Brain,* (2011); 134: 2222-32.

[37] Isono M, Wakabayashi Y, Fujiki MM, Kamide T, Kobayashi H. Sleep cycle in patients in a state of permanent unconsciousness. *Brain injury,* (2002), 16(8):705-12.

[38] Laureys S, Owen AM, Schiff ND. Brain function in coma, vegetative state and related disorders. *Lancet Neurol.,* (2004) 3: 537-46.

[39] Monti MM, Vanhaudenhuyse A, Coleman MR, Boly M, Pickard JD, tshibanda L, owen AM, Laureys S. Willful modulation of brain activity in disorders of consciousness. *N. Eng. J. Med.,* (2010) 362: 579-89.

[40] Bekinschtein TA, Manes FF, Villarreal M, Owen AM, Della Maggiore V. Functional imaging reveals movement preparatory activity in the vegetative state. *Front Hum. Neurosci.,* (2011) 5: 5.

[41] Bardin JC, Fins JJ, Katz DI, Hersh J, Heier LA, Tabelow K, Dyke JP, Ballon DJ, Schiff ND, Voss HU. Dissociation between behavioural and functional magnetic resonance imaging-based evaluations of cognitive function after brain injury. *Brain,* (2011) 134: 769-82.

[42] Schnakers C, Perrin F, Schabus M, Majerus S, Ledoux D, Damas P, Boly M, Vanhaudenhuyse A, Bruno MA, Moonen G, Laureys S. Voluntary brain processing in disorders of consciousness. *Neurology, (*2011) 71;20: 1614-20.

[43] Owen AM, Coleman MR, Boly M, Davis MH, Laureys S, Pickard JD. Detecting awareness in vegetative states. *Science,* (2006) 313: 1402.

In: Chronic Disorders of Consciousness ISBN: 978-1-62618-940-9
Editors: R. S. Calabrò and P. Bramanti

Chapter 4

Neuroimaging Assessment in Low Level Consciousness Patients

***Silvia Marino*[1] *and Nicola De Stefano*[2]**
[1]IRCCS Centro Neurolesi "Bonino-Pulejo", Messina, Italy
[2] Dept. of Neurology, Neurosurgery and Behavioral Sciences, University of Siena, Siena, Italy

Abstract

Consciousness is a multifaceted concept that has two dimensions: arousal, or wakefulness (i.e., level of consciousness), and awareness (i.e., content of consciousness). An accurate and reliable assessment of the arousal and awareness of consciousness in patients with severe brain damage is of greatest importance for the differential diagnosis of low level consciousness patients and for outcome evaluation. Recent studies suggest a number of potential clinical and rehabilitative applications of magnetic resonance (MR) techniques. Although bedside clinical examination remains the criterion standard for establishing diagnosis, MR may provide an adjunctive diagnostic role when behavioural findings are very limited or ambiguous. The future of diagnostic and prognostic assessment of patients with disorders of consciousness envisions a battery of neurobehavioral and neuroimaging techniques (such as structural and functional MR imaging, MR spectroscopy, diffusion tensor imaging, fiber tracking, positron emission tomography) that serve as complementary clinical tools that may help differentiate the effects of underarousal, sensory impairment, motor dysfunction, and cognitive disturbance in the search for potential causes of behavioural unresponsiveness.

1. Introduction

Consciousness is a multifaceted concept that has two dimensions: arousal, or wakefulness (i.e., level of consciousness), and awareness (i.e., content of consciousness). An accurate and reliable assessment of the arousal and awareness of consciousness in patients with severe brain damage is of greatest importance for the differential diagnosis of low level consciousness patients and for outcome evaluation. Following coma, some patients permanently lose all brainstem function (brain death), some progress to "wakeful unawareness" (vegetative state - VS), whereas others recover typically and progress through different stages before fully or partly recovering consciousness (minimally conscious state - MCS). Patients in VS can open their eyes and exhibit basic orienting responses, but show no conscious, purposeful activity. Reflex and other movements are seen, mediated by brainstem, spinal cord, and brainstem-diencephalic arousal systems [1]. VS can occur after patients emerge from an acute catastrophic brain insult causing coma, or can also be seen in degenerative or congenital nervous system disorders. The two common findings are necrosis of the cerebral cortex, thalamus and brainstem (usually after anoxic injury) and diffuse axonal injury (usually after trauma), although other pathological findings can be seen in degenerative and other disorders [2]. The MCS patients do not meet diagnostic criteria for coma or VS because they demonstrate some inconsistent, but clear, evidence of consciousness [2-3]. In the MCS, there is variable impaired function of the cerebral cortex, diencephalons and upper brainstem. This allows occasional conscious behaviours to occur, unlike in VS or coma. Patients may enter the MCS as they emerge from coma or VS, or they can become minimally conscious as a result of acute injury or chronic degenerative diseases. Recent studies suggest a number of potential clinical and rehabilitative applications of magnetic resonance (MR) techniques. Although bedside clinical examination remains the criterion standard for establishing diagnosis, MR may provide an adjunctive diagnostic role when behavioural findings are very limited or ambiguous. The future of diagnostic and prognostic assessment of patients with disorders of consciousness (DOC) envisions a battery of neurobehavioral and neuroimaging techniques (such as structural and functional MR imaging

(MRI and fMRI), MR spectroscopy (MRS), diffusion tensor imaging (DTI), fiber tracking, positron emission tomography (PET)) that serve as complementary clinical tools that may help differentiate the effects of underarousal, sensory impairment, motor dysfunction, and cognitive disturbance in the search for potential causes of behavioural unresponsiveness.

2. Magnetic Resonance and Magnetization Transfer Imaging

The morphological MRI acquisitions usually include non-contrast-enhanced sagittal T1, axial diffusion, axial fluid attenuated inversion recovery (FLAIR), axial T2-SE, coronal T2 sequences and a 3D T1-weighted volume acquisition. FLAIR and T2-SE sequences permit to detect brain edema, contusion, hematoma, herniation, subarachnoid hemorrhage, or hydrocephalus. T2 sequences are useful in detecting hemorrhagic diffuse axonal injuries (DAI). The total number of lesions detected by FLAIR and T2 are shown to be inversely correlated with Glasgow Outcome Scale (GOS) of traumatic coma patients; while the 3D T1 sequence provides an opportunity to evaluate the brain atrophy during the follow up of these patients. A lot of studies performed on traumatic coma patients with conventional MRI showed that lesions of the pons, midbrain, and basal ganglia were predictive of poor outcome especially when they are bilateral. Despite their encouraging results, these studies fail to explain why some patients in VS or with long-term marked cognitive impairments have no or minimal lesions on conventional MRI examination. This raises the question of the lack of specificity and insufficient sensitivity of conventional MR sequences which fail to reveal lesions such as ischemic axonal injuries. Therefore, it is clear that morphological and conventional MRI alone cannot be considered as a reliable tool to assess consciousness disorders severity or to predict their evolution and outcome [4]. Several studies investigated patients in VS and in MCS using non-conventional, quantitative and volumetric MR techniques, useful to provide information about the anatomical patterns, the prognosis and the outcome of these patients. Ammermann et al. [5] have used volumetric analysis of MRI to determine the pattern of lesions in 12 patients with a severe neurological impairment after acute ischemic injury. At the time of scanning, the patients were either in VS or in an early remising state, that is MCS. Lesions were classified as having been present in the gray and/or white matter in four different brain regions (frontal, parietal, temporal, occipital). An additional separate evaluation was performed for the basal ganglia, thalamus, hippocampus, cerebellum, and brainstem. The total clinical follow-up period of all patients from the time of the causative event lasted for at least 5 months. The clinical outcomes were reported according to the Rancho Los Amigos Cognitive Scale (RLACS) as a universal guide to assess a patient's level of functioning. The final RLACS levels were correlated to the MRI lesion size with a Spearman correlation. All patients demonstrated extensive white matter lesions, with the largest lesions observed in the frontal and occipital lobe. A preferential involvement of the white matter located in the periventricular area and in the subcortical regions below the motor and internal temporal cortices was found, in addition to the classifically described lesions of the striatum, motor and occipital cortices. Lesion magnitude showed an association with the severity of the outcome as quantitatively assessed by RLACS. With respect to gray matter lesions, the vulnerability pattern observed included frontal and occipital and in some cases parietal cortical areas, moreover in most cases the thalamus. Additionally, almost all of patients showed lesions of the hippocampus or lesions to the basal ganglia. An association between the extent of the MRI defined lesions located within the white matter and the clinical outcomes of the patients was found. All patients in the most unfavorable class III clinical outcome group (i.e. persistent VS) exhibited white matter lesions exceeding 2/3 of the volume of at least one lobe, most frequently the occipital lobe.

Moreover, Juengling et al. [6] investigated 5 patients in persistent VS due to prolonged cerebral hypoxia of non-traumatic origin, using combined Voxel-Based Morphometry (VBM) of 3D MRI and FDG-PET analysis. In the analysis of the regional distribution of gray matter atrophy, VBM revealed multiple areas of significantly decreased gray matter density at pb0.001, corrected for multiple comparisons. Those were localized in multiple cortical areas, in particular including inferior parietal lobe, superior and medial frontal lobe, paracentral lobule, superior and medial temporal lobe, the cingulum, and the fusiform gyrus. Thalamic changes were limited to small voxel clusters in dorso-medial areas. These structural atrophic changes were compared with the local distribution of functional loss as assessed by regional hypometabolism in the FDG-PET group analysis. At the threshold pb0.001 (corrected for multiple comparisons), PET showed a widespread pattern of hypometabolic areas. In particular, the parietal and frontotemporal cortices, the cuneus/precuneus, the cingulum, the frontal medial and precentral gyrus, and the transverse temporal gyrus were involved, additionally the bilateral thalamus (mainly dorso-medial subnucleus). All changes were, similar to the VBM results, nearly symmetrical. Improved understanding of this complex lesion pattern gained by in vivo group analyses like here might help to provide deeper insights into the general pathoanatomy of patients in the persistent VS.

Using high-resolution T1-weighted magnetic resonance images and a novel approach to shape analysis applied SIENAX software, Fernandez-Espejo et al. [7] investigated thalamic global and regional changes in a sample of patients in a VS or an MCS. They found that total thalamic volume was significantly lower in patients than in healthy volunteers. Shape analysis revealed significant bilateral regional atrophy in the dorso-medial body in patients compared to controls; this atrophy was more widespread in VS than in MCS patients. Lower thalamic volume was significantly correlated with worsening of Disability Rating Scale (DRS) scores. Shape analysis suggested that the dorso-medial nucleus and the internal medullar lamina were the main regions responsible for this correlation. These findings suggest that MCS and VS patients present different patterns of regional thalamic abnormalities. In particular, VS patients showed a more widespread pattern of atrophy than controls, producing differences in global thalamic volume. MCS patients did not show volumetric differences compared to controls, and regionally they showed a less pronounced inward collapse in both the dorsal and ventral areas, with the anterior-ventral body significantly spared. Neuropathological studies have demonstrated that thalamic damage is less common in MCS than in VS patients [8].

Another quantitative RM technique is the Magnetization Transfer Imaging (MTI). The MTI relies on the principle that protons bound in structures exhibit T1 relaxation coupling with protons in the aqueous phase. When an off-resonance saturation pulse is applied, it selectively saturates those protons that are bound in macromolecules. These protons subsequently exchange longitudinal magnetization with free water protons, leading to a reduction in the detected signal intensity [9]. The MTI may provide a quantitative index of the structural integrity of tissue and might be useful to study the outcome of patients with low levels of consciousness.

However, further studies, on larger groups of patients, need to be performed to confirm the usefulness of quantitative MRI in the assessment of the eventual neurological prognosis and outcome of these challenging patients.

3. Functional Magnetic Resonance Imaging and Positron Emission Tomography

At present a diagnosis of VS or MCS is made using prognostic markers from the patient's clinical history supported by detailed neurological and behavioral assessment by a multidisciplinary team over several weeks. However, the behavioral assessment of these patients predominately relies upon the subjective interpretation of observed spontaneous and volitional behavior. A diagnosis of VS is supported if the patient demonstrates no evidence of awareness of self or environment, no evidence of sustained, reproducible, purposeful or voluntary behavioral response to visual, auditory, tactile or noxious stimuli and critically no evidence of language comprehension or expression [10]. In contrast the patient in MCS demonstrates partial preservation of awareness of self and environment, responding intermittently, but reproducibly, to verbal command and therefore demonstrating some degree of basic language comprehension [11].

PET and recently fMRI, by measurement of cerebral metabolism and brain activations in response to sensory stimuli, can provide important MR indices on the presence and location of any residual brain function.

PET is the most sensitive method to image trace amounts of molecules in vivo. Therefore this technique is used to measure in man or in the living animal biochemical and physiological processes in any organ with three dimensional resolution. The last 25 years have seen a rapid and still ongoing development in the production of positron emitters, radiochemical labeling techniques, tomograph technology and image reconstruction algorithms. Because of the possibility to see and measure quantitatively physiological disorders in an early stage, before permanent morphological damage has occurred, which will only then be visible in x-ray or magnetic resonance computer tomography, PET is finally finding its way from a sophisticated research tool into routine clinical diagnosis.

Resting cerebral metabolism derived from quantitative glucose uptake provides an indirect assessment of neuronal activity against which brain states may be compared quantitatively [11]. All previous quantitative [^{18}F] fluorodeoxyglucose–positron emission tomograph (FDG-PET) investigations of VS have correlated the condition with a global reduction of brain metabolic activity: Laureys et al. [12] have assessed regional cerebral glucose metabolism (rCMRGlu) and effective cortical connectivity in four patients in

VS by means of statistical parametric mapping and FDG-PET. Results showed a common pattern of impaired rCMRGlu in the prefrontal, premotor, and parietotemporal association areas and posterior cingulate cortex/precuneus in VS. In a next step, they demonstrated that in VS patients various prefrontal and premotor areas have in common that they were less tightly connected with the posterior cingulate cortex than in normal controls. Schiff et al. 13] have described the first evidence of reciprocal clinical–pathological correlation with regional differences of quantitative cerebral metabolism. They studied five patients in VS with different behavioral features employing FDG-PET, MRI and magnetoencephalographic (MEG) responses to sensory stimulation. Each patient's brain expressed a unique metabolic pattern. The specific patterns of preserved metabolic activity identified in these patients reflect novel evidence of the modular nature of individual functional networks that underlie conscious brain function. In three of the five patients, co-registered PET/MRI correlate islands of relatively preserved brain metabolism with isolated fragments of behavior. Two patients had suffered anoxic injuries and demonstrated marked decreases in overall cerebral metabolism. Two other patients with non-anoxic, multifocal brain injuries demonstrated several isolated brain regions with relatively higher metabolic rates. A single patient who suffered severe injury to the tegmental mesencephalon and paramedian thalamus showed widely preserved cortical metabolism. The variations in cerebral metabolism in chronic VS patients indicate that some cerebral regions can retain partial function in catastrophically injured brains.

fMRI is based on the increase in blood flow to the local vasculature that accompanies neural activity in the brain. This result in a corresponding local reduction in deoxyhemoglobin because the increase in blood flow occurs without an increase of similar magnitude in oxygen extraction [14] [15]. Since deoxyhemoglobin is paramagnetic, it alters the T2 weighted magnetic resonance image signal [16]. Thus, deoxyhemoglobin is sometimes referred to as an endogenous contrast enhancing agent, and serves as the source of the signal for fMRI. Using an appropriate imaging sequence, human cortical functions can be observed without the use of exogenous contrast enhancing agents on a clinical strength (1.5 T) scanner [17, 18].

Functional activity of the brain determined from the magnetic resonance signal has confirmed known anatomically distinct processing areas in the visual cortex [18], the motor cortex, and Broca's area of speech and language-related activities [19-20]. Further, a rapidly emerging body of literature documents corresponding findings between fMRI and conventional electrophysiological techniques to localize specific functions of the human brain [21-22]. Consequently, the number of medical and research centers with fMRI capabilities and investigational programs continues to escalate.

Several fMRI studies in the VS have confirmed the findings of previous PET studies. Di et al. [23] used fMRI to evaluate differences between seven VS and four MCS patients in brain activation occurring in response to the presentation of the patient's own name, spoken by familiar voice (SON-FV). They prospectively studied residual cerebral activation to SON-FV in seven patients with VS and four with MCS. Two patients with VS failed to show any significant cerebral activation. Three patients with VS showed SON-FV induced activation within the primary auditory cortex. Only two of the VS patients, and all four MCS patients, showed activation not only in the primary auditory cortex but also in hierarchically higher-order associative temporal areas.

Three months after fMRI examination, these two VS patients had progressed to the MCS. This study showed that fMRI measurement might be a useful tool for pre-clinically distinguishing MCS-like cognitive processing in some patients behavioural classified as vegetative. Schiff et al. [13] have tested the hypothesis that MCS patients retain active cerebral networks that underlie cognitive function. fMRI was employed to investigate cortical responses in two male adults with severe brain injuries resulting to MCS and in seven healthy volunteers. Three passive stimulation tasks were performed: tactile stimulation, auditory narratives of familiar events presented by a familiar person, and the same auditory passages without language-related content. Results have showed a residual brain activity of cortical systems involved in a potential cognitive and sensory function despite their inability to follow simple instructions or communicate reliably.

In conclusion, results of these studies we analyzed confirm the idea that PET and fMRI activation profiles may constitute usefule adjunctive diagnostic methods when behavioral findings are very limited or ambiguous, helping in differential diagnosis, prognostic assessment and identification of pathophysiological mechanism.

4. Diffusion Tensor Imaging

Diffusion tensor imaging (DTI) is an emerging technique that complements traditional MRI and may be able to provide erstwhile unavailable information about the pathological substrates of DOC. DTI is a modified MRI technique that is sensitive to microscopic, three-dimensional water motion within tissue. In cerebrospinal fluid, water motion is isotropic, i.e., roughly equivalent in all directions. In white matter, however, water diffuses in a highly directional or anisotropic manner. Due to the structure and insulation characteristics of myelinated fibers, water in these white matter bundles is largely restricted to diffusion along the axis of the bundle. DTI can thus be used to calculate two basic properties: the overall amount of diffusion and the anisotropy [24-28]. It is only very recently that DTI has been used to evaluate white matter integrity in patients with DOC. For example, Voss et al. [29] described two patients with traumatic brain injury: one who had remained MCS for 6 years and one who had recovered expressive language after 19 years diagnosed as MCS. In both cases, widespread changes in white matter integrity were observed. Interestingly, however, the increased anisotropy and directionality in the bilateral medial parieto-occipital regions that was observed in the second patient reduced to normal values in a follow-up scan performed 18 months later. This coincided with increased metabolic activity, leading the authors to interpret these observations as evidence of axonal regrowth in this region. Although this is certainly a landmark finding in two high spectrum MCS patients, it remains to be seen whether DTI has any diagnostic or prognostic utility in a broader group of patients with disorders of consciousness. To this end, Tollard et al. [30] and Perlbarg et al. [31] have recently demonstrated that DTI measures in sub-acute severe traumatic brain injury may be a relevant biomarker for predicting the recovery of consciousness at 1 year. However, VS and MCS patients were classified in the same outcome category and potential differences between these two groups were not investigated. Although, in this context DTI has been generally used to address specific clinical problems, the study of white matter integrity in behaviorally defined states has a more basic relevance to understanding the relationship between brain and behavior in both health and disease. For example, in healthy volunteers, DTI techniques have been used recently to examine how structural changes underpin the behavioral changes that are related to learning a complex skill [32]. In a very recent study [33], the integrity of white and grey matter regions was assessed in a group of 25 VS and MCS patients in vivo. In accordance with previous post-mortem work [8, 34] significant changes were observed in the integrity of the tissue in subcortical, thalamic and brainstem regions in the patients when compared to healthy volunteers. The precise location of this damage was not different between the MCS and VS sub-groups, which, again, accords well with previous post-mortem studies. However, an analysis of the MD values within two of these regions of interest (subcortical white matter and thalami), revealed significant differences between the patients meeting the clinical (behavioral) criteria defining VS and those who met the criteria defining MCS. Specifically, the VS patient group exhibited a decrease in the peak height of the histograms derived from the subcortical white matter and the thalami and an increase in the peak width of the thalamic histogram.

In addition, DTI may be a valuable biomarker for the severity of tissue injury and a predictor for outcome. It reveals changes in the WM that are correlated with both acute GCS and Rankin scores at discharge [35]. Significant early reduction of anisotropy was observed in WM structures, in particular in the internal capsule and the corpus callosum, which are the sites most commonly involved by DAI [35]. Moreover, several regions recovered normal values of anisotropy 1 month after the injury [36]. Xu et al. [37] found significant differences in the corpus callosum, internal and external capsule, superior and inferior longitudinal fascicles, and the fornix in TBI patients. They showed that FA and ADC measurements offered superior sensitivity compared to conventional MRI diagnosis of DAI. Salmond et al. [38] reported increased diffusivity in TBI patients at least 6 months after their injury in the cerebellum, frontal, insula, cingulate, parietal, temporal, and occipital lobes. The anisotropy seems to be reduced both in the major WM tracts such as the corpus callosum and the internal and external capsule, and the associative fibers underlying the cortex. DTI has a number of advantages as an imaging biomarker of brain injury: first, it can be used to evaluate brain trauma in an unconscious or sedated patient; second, it could permit the evaluation of responses to treatment even when the clinical scores are inadequate for assessing the patient; third, quantitative DTI measurements are unlikely to be tainted by adverse central nervous system (CNS) effects of hypnotic drugs, unlike clinical scores; and fourth, DTI may be an important alternative marker, as low initial Glasgow Come Scale scores are of limited value in predicting the prognosis [35]. Finally, Perlbarg et al. [31] showed significant FA differences between favorable and unfavorable 1-year outcome groups around four FA tracks: in inferior longitudinal fasciculus, posterior limb of the internal capsule, cerebral peduncle, and posterior corpus callosum.

5. Magnetic Resonance Spectroscopy

Proton MRS (^{1}H-MRS) is a non-invasive imaging technique that enables in vivo quantification of certain neurochemical compounds. Using the same equipment utilized for the conventional MRI, single-voxel ^{1}H-MRS and multi-voxel Imaging (^{1}H-MRSI) or Chemical Shift Imaging (CSI) provide metabolic information on brain damage that may not be visible with the conventional structural imaging methods. Then ^{1}H-MRS, added to traditional MRI, offers the possibility to study the brain activity combining information on structure and function.

Classically, the exploration of DOC is performed on 1,5 or 3 Tesla MR scanners and at intermediate or long echo time (TE) (135-288 ms). Long TE ^{1}H-MRS detects the signal arising from four metabolites: N-acetyl-aspartate containing compounds (NAA), choline-containing compounds (Cho), creatine + phosphocreatine (Cre) and lactate (Lac). Short TE ^{1}H-MRS identifies peaks from mobile lipids, Lac, alanine, NAA, Glutamate/Glutamine (Glx), γ-aminobutyric acid, Cre, Cho, myo-inositol, and scyllo-inositol (Figure 1).

NAA, which resonates at 2.02 parts per million (ppm), represents the largest proton metabolic concentration in the human brain after water. Indeed the concentration of NAA reaches on the order of 10 μmol/g. NAA is widely interpreted as a neuronal marker and implicated in several neuronal processes, mitochondrial functioning and osmoregulation. NAA synthesis occurs in mitochondria and requires acetyl-CoA and L-aspartic acid as substrates. NAA has been proposed to serve as a mitochondrial shuttle of acetyl-CoA used for fatty acid synthesis. Its peak decreases when there is neuron suffering or loss. The Cho peak (3.2 ppm) represents a combination of several choline-containing compounds, including free Cho, phosphorylcholine and glycerophosphorylcholine, and to a small extent acetylcholine. Free Cho acts as a precursor to acetylcholine, while glycerophosphorylcholine is a product of breakdown of membrane phosphatidylcholine and acts as an osmoregulator. Its peak increases when there is greater membrane tournover, cell proliferation or inflammatory process. The peak of Cre at 3.03 ppm represents total creatine and phosphocreatine supplies phosphate for conversion of ADP to ATP in creatine kinase reaction. Indeed these metabolites buffer the energy use and energy storage of cells. The level of total Cre mainly remains constant in many neuronal diseases. Thus, total Cre is often used as an internal reference (i.e., a denominator in metabolite signal ratio). The Lac (1.3 ppm) is an end product of anaerobic glycolysis, thus increase in Lac concentrations often serves as an index of altered oxidative metabolism, i.e., in ischemia, hypoxia, and cancer. Increases of Lac in the brain are often accompanied by decreased intracellular pH and high-energy phosphates. The proposed role of Lac is a source of energy for neurons and the transport of Lac plays an essential role in the concept of metabolic coupling between neurons and glia. Glutamate (Glu) is the highest excitatory neurotransmitter in concentration in the CNS. Its peak increases when neuronal and astrocytic activation impairs mitochondrial function and energy utilization. Indeed this process impairs Glu transport and its following enhancement is associated to cellular toxicity.

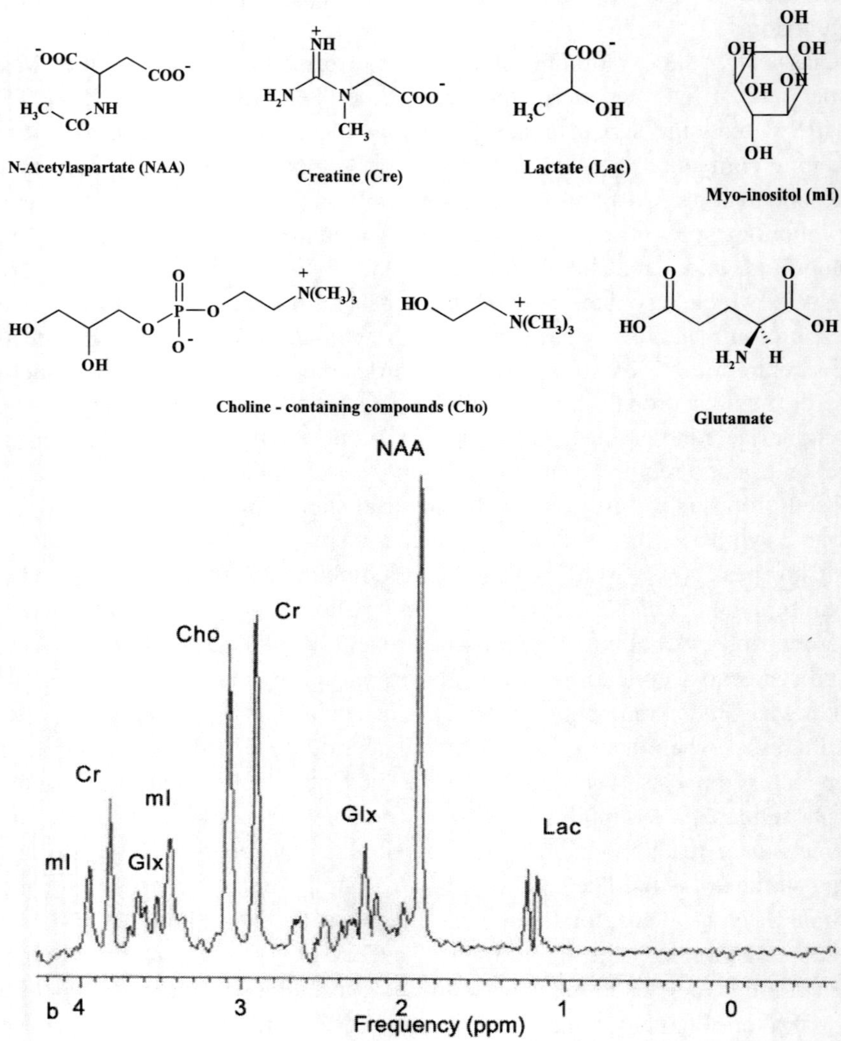

Figure 1. Chemical structure and spectrum of main cerebral metabolites detected by ^{1}H-MRS.

^{1}H-MRS has been used for at least 15 years in the exploration of patients with altered consciousness, both to investigate the mechanisms of vigilance and to predict the possibilities of regaining consciousness.

Predicting outcome of patients with DOC is an integral part of clinical care, facilitating medical decision making and therapeutic intervention. Current neurological and neurophysiological methods do not enable prediction of outcome of these patients in early stages. Although conventional neuroimaging can provide important information for acute clinical management, its prognostic value is limited, particularly at early stage of injury resolution, owing to its poor sensitivity.

Several studies present in literature have demonstrated the value of ^{1}H-MRS as an accurate tool to predict patient's clinical outcome. Indeed many investigators have shown that correlation exists between metabolite changes and outcome of patients with DOC.

Previous studies using single-voxel technique have shown in brain-injured subjects a significant correlation between unfavorable outcome and reduction of marker NAA in occipitoparietal white and gray matter (WM and GM) [39-42], frontal WM [43], parietal WM [44], brainstem [45], splenium of the corpus callosum [9, 46], and thalamus [47], increase in choline a marker for cell membrane disruption in frontal WM [43] and occipitoparietal WM and GM [39, 46, 41-42] and increase in Glx in occipital GM and parietal WM [44].

In particular, NAA levels seem to discriminate patients who recovered from coma from those who died or remained in persistent VS [48]. Uzan et al. [47] carried out a thalamic proton MRS in patients in VS resulting from severe TBI. They found that NAA/Cr ratios were able to differentiate patients in VS who recovered awareness from those who remained in persistent VS. However, this alteration was not found in the thalamus of patients in VS resulting from mild TBI [49].

In some studies has been shown that the combination of imaging techniques may be useful to predict the long-term neurological outcome. A ^{1}H-MRS study in the pons allowed separating of patients who recovered from patients with severe neurological impairment, death or in VS. In addition, ^{1}H-MRS metabolic alterations were not correlated with anatomical MRI lesions, suggesting that these two techniques are strongly complementarity [45]. Tollard et al. [30] reported the first study on patients with TBI based on a combined quantitative analysis of ^{1}H-MRS and DTI. This combined analysis was 97% specific for predicting an unfavorable outcome after 1 year, compared with 85% for DTI and 75% for ^{1}H-MRS. Similarly, sensitivity was better with the combined analysis (86%) than with either DTI (79%) or MRS (75%).

To study metabolite changes from a wider area of the brain, with the advantage of identifying more anatomical and functional details, a few investigators have used ^{1}H-MRSI [50-54]. This technique has an advantage over single-voxel ^{1}H-MRS because generates individual spectra from multiple voxels at the same time. Also ^{1}H-MRSI studies have highlighted close correlation between metabolite alterations and potential recovery.

Neurometabolite concentrations obtained soon after injury may be useful for predicting individual outcome. The decrease of NAA and the increase of Lac, seen by Marino et al. [51] early after brain injury, were correlated with GOS score. Then these ^{1}H-MRS data may be, at this stage, a reliable index of injury severity and disease outcome.

However it is need to note that ^{1}H-MRS studies in patients with DOC are heterogeneous in terms of patient nature, injury types, time from cerebral damage, voxel location, methods and timing outcome assessment. In addition, in many studies the metabolite concentrations were expressed in terms of semiquantitative ratios. The assumption that the concentration of Cr as reference metabolite remains constant may be incorrect, especially in acute conditions. It is therefore advisable to obtain concentration expressed in standard units by applying absolute quantification. Some studies have expressed metabolite concentrations in term of absolute quantification [39, 44].

Data reported so far demonstrate that MRS measure have the potential to provide new and important biological brain markers able to predict clinical outcome, helping in the therapeutic interventions, clinical and rehabilitative management of these patients, as well as to assist with family education.

Conclusion

Patients with severe brain damage who are unable to communicate present several ethical concerns. Foremost is the concern that diagnostic and prognostic accuracy is certain, as treatment decisions typically include the possibility of withdrawal of life-support. Although imaging techniques have the potential to improve both diagnostic and prognostic accuracy, careful and repeated neurological assessment by a trained examiner remains best practice. Accurate clinical assessments of patients in these conditions must be obtained before they undergo neuroimaging. Moreover, in reports of neuroimaging studies, all relevant clinical details must be available for comparisons between studies.

Ethical concerns are commonly raised about the participation of patients with severe brain damage in neuroimaging studies. By definition, unconscious or minimally conscious patients cannot give informed consent to participate in clinical research and written approval must typically be obtained from family or legal representatives depending on governmental and hospital guidelines. Nonetheless, researchers studying these patients have been refused grants, ethics committee approval, and research publication; these decisions tend to be made on the basis that studies of patients who cannot provide consent are unethical. We prefer an ethical framework that balances access to research with medical advances alongside protection for defenseless patients. Severe brain damage represents an immense social and economic problem that warrants further research. Unconscious, minimally conscious, and locked-in patients deserve special procedural protections. However, it is important to stress that they are also at risk of being denied therapy that may be life-saving if clinical research cannot be done on these patient groups.

Patients who are in coma, VS, MCS, or locked-in syndrome present unique problems for diagnosis, prognosis, treatment, and everyday management. At the patient's bedside, assessment of cognitive function is difficult because voluntary movements may be very small, inconsistent, and easily exhausted. Functional neuroimaging will never replace the clinical assessment of patients with altered states of consciousness. Nevertheless, using population norms it can provide an objective measure of the regional distribution of cerebral activity at rest and under various conditions of stimulation. The quantification of brain activity differentiates patients who sometimes only differ by a brief and small movement of a finger. In our opinion, PET, MRS and fMRI will increase substantially our understanding of patients with severe brain damage.

References

[1] Laureys, S; Owen, AM & Schiff, ND. Brain function in coma, vegetative state, and related disorders. *Lancet Neurology* 2004;. 3 (9):537-546.

[2] Laureys, S & Boly, M The changing spectrum of coma. *Nature Clinical Practice Neurology* 2008; 10(10):544-546.

[3] Giacino, JY; Ashwal, S; Childs, N; Cranford, R; Jennett, B et al. The minimally conscious state: definition and diagnostic criteria. *Neurology* 2002; 58(3): 349-353.

[4] Tshibanda, L; Vanhaudenhuyse, A; Boly, M; Soddu, A; Bruno, MA et al. Neuroimaging after coma. *Neuroradiology* 2010; 52(1):15-24.

[5] Ammermann, H; Kassube, KJ; Lotze, M; Gut, E; Kaps, M et al. MRI brain lesion patterns in patients in anoxia-induced vegetative state. *Journal of the Neurological Sciences* 2007; 260(1-2); 65-70.

[6] Juengling, FD; Kassubek, J; Huppertz, H; Krause, T & Els, T. Separating functional and structural damage in persistent vegetative state using combined voxel-based analysis of 3-D MRI and FDG-PET. *Journal of the Neurological Sciences* 2005; 228(2):179–184.

[7] Fernandez-Espejo, D; Junque, C; Bernabeu, M; Roig-Rovira, T; Vendrell P. et al. Reductions of thalamic volume and regional shape changes in the vegetative and the minimally conscious states. *Journal of Neurotrauma* 2010; 27(7): 1187–1193.

[8] Jennett, B; Adams, JH; Murray, LS & Graham, DI. Neuropathology in vegetative and severely disabled patients after head injury. *Neurology* 2001; 56(4): 486–490.

[9] Sinson, G; Bagley, LJ; Cecil, KM; Torchia, M; McGowan, JC et al. Magnetization transfer imaging and proton MR spectroscopy in the evaluation of axonal injury: Correlation with clinical outcome after traumatic brain injury. *American Journal of Neuroradiology* 2001; 22(1): 143-151.

[10] Medical aspects of the persistent vegetative state (1). The Multi-Society Task Force on PVS. *New England Journal Medicine* 1994; 330(21): 1499-508.

[11] Levy, DE; Sidtis, JJ; Rottenberg, DA; Jarden, JO; Strother, SC; Dhawan, V et al. Differences in cerebral blood flow and glucose utilization in vegetative versus locked-in patients. *Annals of Neurology* 1897; 22(6)

[12] Laureys, S; Goldman, S; Phillips. C; Van Bogaert, P; Aerts, J et al. Impaired effective cortical connectivity in vegetative state: preliminary investigation using PET. *Neuroimage* 1999; 9(4): 377–82.

[13] Schiff, ND; Rodriguez-Moreno, D; Kamal, A; Kim, KHS; Giacino, JT et al. fMRI reveals large-scale network activation in minimally conscious patients. *Neurology* 2005; .64(11):1843-1843-a.

[14] Roy, C & Sherrington, C. On the regulation of the blood-supply of the brain. *The Journal of Physiology* 1890; 11(2): 85-108.

[15] Fox, PT & Raichle, ME. Stimulus rate determines regional brain blood flow in striate cortex. *Annals of Neurology* 1985; 17(3):303-305.

[16] Ogawa, S; Lee, TM; Nayak, AS & Glynn, P. Oxygenation-sensitive contrast in magnetic resonance image of rodent brain at high magnetic fields. *Magnetic Resonance in Medicine* 1990;.14(1): 68-78.

[17] Bandettini, PA; Jesmanowicz, A; Wong, EC & Hyde, JS. Processing strategies for time-course data sets in functional MRI of the human brain. *Magnetic Resonance in Medicine* 1993; 30(2): 161-173.

[18] Schneider, W; Noll, DC & Cohen, JD. Functional topographic mapping of the cortical ribbon in human vision with conventional MRI scanners. *Nature* 1993; 365(6442): 150-152.

[19] Hinke, RM; Hu, X; Stillman, Kim, SG; Merkle, H et al. Functional magnetic resonance imaging of Broca's area during internal speech. *NeuroReport* 1993; 4(6): 675-678

[20] Kim, K; Hirsch, J; DeLaPaz, RL; Relkin, N & Lee, KM Comparison of cortical areas activated by primary and secondary languages in human brain using functional magnetic resonance imaging (fMRI). Abstracts: Society of Neuroscience 1993; 21(3):1763.

[21] Atlas, SW; Howard, IIRS; Maldijian, J; Alsop, D; Detre, JA et al. Functional magnetic resonance imaging of regional brain activity in patients with intracerebral gliomas: findings and implications for clinical management. *Neurosurgery* 1996; 38(2): 329-338.

[22] George, JS; Aine, CJ; Mosher, JC; Schmidt, MD; Ranken, DM et al. Mapping function in the human brain with magneto encephalography, anatomical magnetic resonance imaging, and functional magnetic resonance imaging. *Journal of Clinical Neurophysiology* 1995; 12(5): 406-429.

[23] Di, H; Yu, SM; Weng, XC; Laureys, S; Yu, D et al. Cerebral response to patient's own name in the vegetative and minimally conscious state. *Neurology* 2007; 68(12):895–899.

[24] Douaud, G; Smith, S; Jenkinson, M; Behrens, T; Johansen-Berg, H et al. Anatomically related grey and whitematter abnormalities in adolescent-onset schizophrenia. *Brain* 2007; 130(9):2375-2386.

[25] Benson, R; Meda, S; Vasudevan, S; Kou, Z; Govindarajan, KA *et al.* Global white matter analysis of diffusion tensor images is predictive of injury severity in traumatic brain injury. *Journal of Neurotrauma* 2007; 24(3): 446-459.

[26] Kraus, MF; Susmaras, T; Caughlin, BP; Walker, CJ; Sweeney JA et al. White matter integrity and cognition in chronic traumatic brain injury: A diffusion tensor imaging study. *Brain* 2007; 130(19): 250-2519.

[27] Ringman, JM; O'Neill, J; Geschwind, D; Medina, L; Apostolova, LG et al. Diffusion tensor imaging in preclinical and presymptomatic carriers of familial Alzheimer's disease mutations. *Brain* 2007; 130(7): 1767-1776.

[28] O'Sullivan, M; Singhal, S; Charlton, R. & Markus HS. Diffusion tensor imaging of thalamus correlates with cognition in CADASIL without dementia. *Neurology* 2004; 62(5): 702-707.

[29] Voss, HU; Uluc, AM; Dyke, JP; Watts, R; Kobylarz, RJ et al. Possible axonal regrowth in late recovery from the minimally conscious state. *The Journal of Clinical Investigation* 2006; 116(7): 2005-2011.

[30] Tollard, E; Galanaud, D; Perlbarg, V; Sanchez-Pena, P; Le Fur, Y. et al. Experience of diffusion tensor imaging and 1H spectroscopy for outcome prediction in severe traumatic brain injury: Preliminary results. *Critical Care Medicine* 2009; 37(4): 1448-1455.

[31] Perlbarg, V; Puybasset, L; Tollard, E; Lehericy, S; Benali, H. et al. Relation between brain lesion location and clinical outcome in patients with severe traumatic brain injury: A diffusion tensor imaging study using voxel-based approaches. *Human Brain Mapping* 2009; 30(12): 3924-3933.

[32] Scholz, J; Klein, MC; Behrens, TE & Johansen-Berg H. Training induces changes in white-matter architecture. *Nature Neuroscience* 2009; 12(11): 1370-1371.

[33] Fernández-Espejo, D; Bekinschtein, T; Monti, MM; Pickard, JD; Junque, et al. Diffusion weighted imaging distinguishes the vegetative state from the minimally conscious state. *Neuroimage* 2011; 51(1):103-112.

[34] Adams, JH; Jennett, B; McLellan, DR; Murray, LS & Graham DI. The neuropathology of the vegetative state after head injury. *Journal* of Clinical *Pathology* 1999; 52(11): 804-806.

[35] Huisman, TA; Schwamm, LH; Schaefer, PW; Koroshetz, WJ; Shetty-Alva, N et al. Diffusion tensor imaging as potential biomarker of white matter injury in diffuse axonal injury. *American Journal of Neuroradiology* 2004; 25(3): 370-376.

[36] Arfanakis, K; Haughton, VM; Carew, JD; Rogers, BP; Dempsey, RJ et al. Diffusion tensor MR imaging in diffuse axonal injury. *American Journal of Neuroradiology* 2002; 23(5): 794-802.

[37] Xu, J; Rasmussen, IA; Lagopoulos, J & Håberg, A. Diffuse axonal injury in severe traumatic brain injury visualized using high-resolution diffusion tensor imaging. *Journal of Neurotrauma* 2007; 24(5): 753-765.

[38] Salmond, CH; Menon, DK; Chatfield, DA; Williams, GB; Pena, A et al. Diffusion tensor imaging in chronic head injury survivors: Correlations with learning and memory indices. *Neuroimage* 2006; 29(1):117-124.

[39] Brooks, WM; Stidley, CA; Petropoulos, H; Jung, RE; Weers, DC et al. Metabolic and cognitive to human traumatic brain injury: a quantitative proton magnetic resonance study. *Journal of Neurotrauma* 2000;17(8):629-640.

[40] Friedman, SD; Brooks, WM; Jung, RE; Chiulli, SJ; Sloan, JH et al. Quantitative proton MRS predicts outcome after traumatic brain injury. *Neurology* 1999; 52(7): 1384-1391.

[41] Ross, BD; Ernst, T; Kreis, R; Haseler, LJ; Bayer, S et al. 1H MRS in acute traumatic brain injury. *Journal of Magnetic Resonance Imaging* 1998; 8(4): 829-840.

[42] Yoon, SJ; Lee, JH; Kim, ST & Chun MH. Evaluation of traumatic brain injured patients in correlation with functional status by localized 1H MR spectroscopy. *Clinical Rehabilitation* 2005; 19(2):209-215.

[43] Garnett, MR; Blamire, AM; Corkill, RG; Cadoux-Hudson, TAD; Rajagopalan B et al. Early proton magnetic resonance spectroscopy in normal-appearing brain correlates with outcome in patients following traumatic brain injury. *Brain* 2000; 123(10): 2046-2054.

[44] Shutter, L; Tong, KA & Holshouser BA. Proton MRS in acute traumatic brain injury: Role for glutamate/glutamine and choline for outcome prediction. *Journal of Neurotrauma* 2004; 21(12): 1693-1705.

[45] Carpentier, A; Galanaud, D; Puybasset, L; Mullar, JC; Lescot, T. et al. Early morphologic and spectroscopic magnetic resonance in severe traumatic brain injuries can detect "invisible brain stem damage" and predict "vegetative states". *Journal of Neurotrauma* 2006, 23(5):674-685.

[46] Cecil, KM; Hills, EC; Sandel, ME; Smith, DH; McIntosh, TK et al. Proton magnetic resonance spectroscopy for detection of axonal injury in the splenium of the corpus callosum of brain-injured patients. *Journal of Neurosurgery* 1998; 88(5):795-801.

[47] Uzan, M; Albayram, S; Dashti, SG; Aydin, S; Hanci, M et al. Thalamic proton magnetic resonance spectroscopy in vegetative state induced by traumatic brain injury. *Journal of Neurolology, Neurosurgery and Psychiatry* 2003; 74(1): 33-38.

[48] Ricci, R; Barbarella, G; Musi, P; Boldrini, P; Trevisan, C et al. Localised proton MR spectroscopy of brain metabolism changes in vegetative patients. *Neuroradiology* 1997; 39(5): 313-319.

[49] Kirov, I.; Fleysher, L.; Babb, JS; Silver, JM.; Grossman, RI et al. Characterizing ''mild'' in traumatic brain injury with proton MR spectroscopy in the thalamus: initial findings. *Brain Injury* 2007; 21(.11):1147-1154.

[50] Holshouser, BA.; Tong, KA; Ashwal, S; Oyoyo, U; Ghamsary, M et al. Prospective longitudinal proton magnetic resonance spectroscopic imaging in adult traumatic brain injury. *Journal of Magnetic Resonance Imaging* 2006; 24(1): 33-40.

[51] Marino, S; Zei, E; Battaglini, M; Vittori, C; Buscalferri, A et al. Acute metabolic brain changes following traumatic brain injury and their relevance to clinical severity and outcome. *Journal of Neurology, Neurosurgery and Psychiatry* 2007, 78,(5): 501-507.

[52] Shutter, L; Tong, KA; Lee, A & Holshouser, BA. Prognostic role of proton magnetic resonance spectroscopy in acute traumatic brain injury. *Journal of Head Trauma Rehabilitation* 2006; 21(4): 334-349.

[53] Signoretti, S; Marmarou, A; Fatouros, P; Hoyle, R.; Beaumont, A et al. Application of chemical shift imaging for measurement of NAA in head injured patients. *Acta Neurochirurgica. Supplement* 2002; : 373-375.

[54] Signoretti, S; Marmarou, A; Aygok, GA; Fatouros, PP; Portella, G. et al. Assessment of mitochondrial impairment in traumatic brain injury using high-resolution proton magnetic resonance spectroscopy. *Journal of Neurosurgery* 2008; 108(1): 42-52.

In: Chronic Disorders of Consciousness ISBN: 978-1-62618-940-9
Editors: R. S. Calabrò and P. Bramanti

Chapter 5

New Diagnostic Tools and Markers of Recovery in Consciousness Disorders

Silvia Marino, Placido Bramanti, Marina Barresi, Debora Celi, Rosella Ciurleo, Simona De Salvo, Valeria Foti, Sabrina Giacoppo and Rocco Salvatore Calabrò
IRCCS Centro Neurolesi Bonino Pulejo, Messina, Italy

Abstract

In recent years, advances in medicine, intensive care and diagnostic imaging procedures, have increased the number of patients who survive acute brain injuries and that are in a vegetative or minimally conscious state. However there are not sufficient findings to evaluate and quantify the severity of the initial and secondary processes destructive and therefore there are not effective therapeutic measures to effectively predict the outcome. It has been argued that diagnostic approaches such as biochemical and neurophysiological markers would provide evidence for pathology progression and help guide therapy development. To date, studies have not found a marker matching such an ideal .For this reason, neuron-specific enolase, S100 calcium binding protein B, myelin basic protein, creatine kinase brain isoenzyme, glial fibrilary acidic protein, plasma desoxyribonucleic acid, brain-derived neurotrophic factor and ubiquitin carboxy-terminal hydrolase-L1 have been proposed as potential markers for cell damage in the central nervous system, to improve the diagnosis and the evaluation of outcome after injury. Also, the neurophysiological examinations, such as Electroencephalogram (EEG), Evoked Potentials (EPs), including olfactory event-related potentials, and Electroencephalogram, in association with functional Magnetic Resonance Imaging, allow monitoring of clinical conditions and they can be a valuable support for the outcome evaluation and clinical differentiation between vegetative or minimally conscious state.

Introdution

In recent years, advances in medicine, intensive care and diagnostic imaging procedures, have increased the number of patients who survive acute brain injuries. Although some of these patients go on to make recovery, many do not and remain in one of several states now known collectively as "disorders of consciousness". These include the vegetative state, the minimally conscious state and coma [1].

However there are not sufficient findings to evaluate and quantify the severity of the initial and secondary processes destructive and therefore there are not effective therapeutic measures to effectively predict the outcome. It has been argued that diagnostic approaches such as biochemical and neurophysiological markers would provide evidence for pathology progression and help guide therapy development.

1. Biochemical Markers in Consciousness Disorders

Recently there has been increasing interest in biochemical markers of brain damage, caused by traumatic and vascular events. If one could design the ideal biochemical marker for neuronal injury, it would be released exclusively by damaged neurons, readily cross the Blood–Brain Barrier (BBB), and be easily detected by laboratory tests. This would result in a marker with high sensitivity and specificity, ease and safety of collection, and a correlation between quantitative analysis and extent of injury. To date, studies have not found a marker matching such an ideal.

Therefore many proteins synthesized in the astroglia cells or in the neurons, such as neuron-specific enolase, S100 calcium binding protein B, myelin basic protein, creatine kinase brain isoenzyme, glial fibrilary acidic protein, plasma desoxyribonucleic acid, brain-derived neurotrophic factor and ubiquitin carboxy-terminal hydrolase-L1, have been proposed as potential markers for cell damage in Central Nervous System (CNS). In specific, the levels of these proteins increase following brain injury and are found in increasing concentrations in the Cerebrospinal Fluid (CSF) depending on the injury magnitude, and can also be found in blood stream because of a compromised BBB [2].

S100 Calcium Binding Protein B

S100 calcium binding protein B (S100B) has been investigated as a possible marker of brain injury in recent years. The S100B protein belongs to a multigenic family of low molecular weight (9-13 kDa) calcium-binding S100 proteins [3, 4]. They have been named "S100" because of their biochemical property of remaining soluble after precipitation with 100% ammonium sulfate [5]. Intracellularly, this protein is involved in signal transduction via the inhibition of protein phosphorylation, regulation of enzyme activity, by affecting the calcium homeostasis. Moreover, S100B is functionally involved in the regulation cell morphology by interaction with elements of the cytoplasmatic cytoskeleton. It is most abundant in the glial cells of the CNS, mainly in astrocytes, however, it may also be found in other tissues such as muscle and fat and constitutes 1-1.5 μg/mg of soluble protein [6].

There is growing evidence that liberation of S100B is also induced by other extracerebral events such as stressful situations. It is known to cross the BBB in measurable quantities after head injuries, even those classified as mild. Maximum concentration peak occurs after 20 minutes, it is metabolized by the kidney and excreted in urine (half-life of about 30 to 113 minutes) [7]. It may be measured in arterial or venous blood, is not affected by hemolysis and remains stable for hours. Elevated serum levels of S100B have been documented in patients suffering from different types of brain damage, including stroke, minor and severe head injury, and brain damage associated with subarachnoidal hemorrhage and at postoperative from cardiac surgery, if followed by neurological complications. Increases in serum S100B concentrations represent irreversible astrocyte injury or death. It is not clear if protein release depends on irreversible cell damage or if it can take place after less severe injury. There is evidence that secretion of S100B by astrocytes might be an active process [8].

Clinically, numerous reports have found a correlation between serum S100B level and the outcome of patients; indeed these indicate, that S100B is released after brain insults and serum levels are positively correlated with the degree of injury and negatively correlated with outcome [9, 10].

Neuron-Specific Enolase

Neuron-specific enolase (NSE) is a glycolytic enzymes, that is functionally active as a heterodimer assembled from α, β and γ subunits, predominantly located in neurons and neuroectodermal cells. The role of NSE is not fully understood, but is probably involved in increasing neuronal chloride levels during the onset of neural activity. The molecular weight of NSE is 78 kDa, and its biological half-life is probably longer than 20 hours, and serum levels of >10 microg/l are considered to be pathological. NSE is the only marker that directly assesses functional damage to neurons. It is passively released by cell destruction and its increased concentrations after neuron damage may be measured in peripheral blood or liquor [11]. Increased levels of NSE were found in blood and liquor of patients with stroke, intracerebral hemorrhage and after cardiopulmonary resuscitation [12]. Also, high serum concentrations of NSE are found in traumatic brain injury (TBI) correlating with injury severity: in severe TBI, NSE correlates with clinical outcome [13, 14]. Normally, it increases in the first 12 hours after trauma and then decreases in the hours and days; its half-life is approximately 24 hours. Secondary increases may take place in patients who evolve to fatal outcome. Although NSE initially appeared to be a promising marker because of a number of theoretical advantages, mainly it is a marker of neurons rather than glial cells and has a high specificity for the brain [15]. One of the main problems associated with use as marker of brain damage is hemolysis. Erythrocytes contain a large amount of NSE and hemolysis may, therefore, cause a marked increase of NSE in the blood. Furthermore, an increase in NSE has been documented clinically in patients with multiple traumas but without head injury, and experimentally in rats with ischaemic injury to abdominal organs [16].

Mielin Basic Protein

Mielin basic protein (MBP) is one of the two most abundant proteins in myelin in the CNS, with a molecular weight of 18.5 kDa.

Studies of MBP concentrations indicate the serum level of MBP is increased after severe traumatic brain injury, demyelinating diseases and in patients with intracranial hemorrhage [8,17,18].

It was also found that mean MBP levels between two and six days after injury were significantly higher in patients with poor outcomes compared with those with good outcomes.

In a small series of 25 patients, Yamazaki et al. similarly found that MBP levels were significantly higher in patients who died compared with those who survived [19].

Glial Fibrillary Acidic Protein

Glial fibrillary acidic protein (GFAP) is a monomeric intermediate filament protein of astrocytes and is only found in glial cells of the CNS. Thus, in contrast to other markers, it is thought to be strictly specific to damage of tissue in CNS. Significantly increased levels of GFAP are observed in CSF as a consequence of acute CNS injury and modestly increased levels of GFAP are seen in patients with astrogliosis. Also, high levels of GFAP are found in blood after stroke, correlating to a functional prognosis [20].

After brain damage proteins like GFAP are released from injured brain cells and appear in the systemic circulation probably directly via passage through a disturbed BBB. Serum analysis has clinical advantages over CSF measurement and CSF sampling is often contraindicated in patients with severe traumatic brain injury. Missler reported a method for serum-GFAP determinations and preliminary results from patients with severe head trauma in 1999 [21]. They concluded that measurement of GFAP concentrations in blood appeared to have the possibility to identify acute CNS damage.

Pelinka and Nylen documented a correlation between serum GFAP levels, severity of injury and outcome [16, 22]. More importantly, it was found that GFAP levels remained normal in multiple trauma patients without head injuries [16].

Plasma Desoxyribonucleic Acid

Plasma cell-free desoxyribonucleic acid (DNA) has been investigated as a potential marker for injury severity and prediction of fatal outcomes [23]. Significantly increased levels of plasma DNA have also been found in trauma patients and they were correlated with a level of severity and development of post-trauma complications [24, 25]. Mechanisms by which free circulating DNA increases after trauma have not been established. High concentrations observed very early after injury, suggest that the extra-cell DNA originates from tissue damage, necrosis, while apoptosis mechanisms may contribute to persistent increases, in addition to a DNA clearance, probably impaired by commitment of the responsible organs due to systemic inflammation [25]. Higher DNA concentrations were significantly associated with fatal outcome when plasma levels were assessed 24 hours after study entry [23].

Brain-Derived Neurotrophic Factor

Brain-derived neurotrophic factor (BDNF) is a member of the neurotrophin family and is the most widely distributed trophic factor in the brain [26]. It participates in neuronal growth, maintenance, and in different aspects of activity-dependent synaptic physiology by acting across different spatial and temporal domains [27]. In the adult CNS, BDNF is expressed by multiple cell types including neurons and glia [28]. Previous research has established that neuronal activity regulates *bdnf* transcription, where BDNF mRNA and protein are subsequently transported into neuronal processes followed by activity-dependent secretion of BDNF. Increased expression of BDNF in the CNS, in response to various stimuli, suggests a neuroprotector role for this neurotrophine. Especially, inflammatory responses seem to be involved in the increased expression of BDNF [29].

Ubiquitin Carboxy-Terminal Hydrolase-L1

Ubiquitin carboxy-terminal hydrolase-L1 (UCH-L1) is also known as neuronal-specific protein gene product 9.5 (PGP9.5) and was previously used a histologic marker for neurons because of its high abundance and specific expression in neurons. UCH-L1 is present in almost all neurons and averages 1-5% of total soluble brain protein [30]. UCH-L1 is involved in the addition or removal of ubiquitin from abnormal proteins including misfolded proteins and proteins damaged by oxidation or denatured by other means destined for proteasomal degradation and has previously been used as a neuronal cell soma marker.

Rapid appearance of a biomarker in biologic material is essential to its clinical utility. Additionally, UCH-L1 remained elevated in patients who experienced post-injury complications and may have added value in the management of patients in the intensive care unit [31].

Creatine Kinase Brain Isoenzyme

Creatine kinase brain isoenzyme (CKBB) is an isoform of creatine-kinase found in the CNS. The brain has many isoforms of CKBB and mitochondrial-CK but lacks creatine kinase muscle b (CKMB) and creatine kinase muscle (CKMM), respectively found in the cardiac and skeletal muscles [32]. The molecular mass of CKBB is 40 to 53 kDa [11]. CKBB is located in the astrocytes and is released when there is anatomical injury in the brain tissue. Its serum levels increase during the first hours after trauma and drop quickly, unless there is a continued enzyme release. Various situations of brain injury such as cardiac arrest and subarachnoid hemorrhage may lead to release of CKBB in the liquor [32].

Statistically significant elevations of CK-BB isoenzyme in the serum were observed in patients with acute cerebrovascular accidents and in those with seizures and a prolonged alteration in level of consciousness. Therefore, statistically significant elevations of CK-BB isoenzyme in the cerebrospinal fluid were also found in patients who had suffered acute cerebrovascular accidents. Some patients with CNS infections, acute demyelinating disease, certain drug overdoses, head trauma, and complex migraine also had elevations of CK-BB isoenzyme in serum and in CSF. In two patients with elevations of CK-BB isoenzyme, the elevations in serum occurred later than those detected in the cerebrospinal fluid [33]. However, the specificity and sensitivity of determination in serum as a predictor of CKKB brain lesions remains controversial [8].

2. Neurophysiological Approach to Evaluate Markers of Recovery in Consciousness Disorders

The neurophysiological approach to patient with consciousness disorders consent the recording of electrical activities of both central and peripheral nervous system and provides a functional assessment.

The neurophysiological examinations, such as Electroencephalogram (EEG), Evoked Potentials (EPs), Electroencephalogram in association with functional Magnetic Resonance, allow monitoring of clinical conditions and are a support for the outcome evaluation and clinical differentiation between VS and MCS.

EEG is useful as an objective electrophysiologic assay of cortical function in patients who do not respond to normal sensory stimuli.

EEG shows an important general slowing of the electrical brain activity of VS patients.

The electroencephalogram in patients in a persistent vegetative state (PVS) has shown a spectrum of abnormalities with changes during the wake-sleep cycle. Patterns have included delta and theta activity and spindle and alpha-like rhythms, but they are more diffusely distributed than in the typical posterior regions and are not reactive to sound, pain, and light stimuli [34]. Clinical recovery from the vegetative state may be paralleled by diminished delta and theta activity and reappearance of reactive alpha rhythm. Indeed, Babiloni et al. has observed that occipital source power in the alpha band of resting EEG, when calculated with low-resolution electromagnetic tomography (LORETA), is correlated with recovery outcome at 3-months follow-up in a group of VS patients; those who made a behavioral recovery had higher resting alpha band power than those who did not make a significant recovery. These results suggest that cortical sources of resting alpha rhythms are related to the outcome of PVS patients at 3-months follow up.

Future perspective research should confirm that cortical sources of resting alpha rhythms might predict recovery in PVS patients [35].

EEG reactivity can manifest itself, not only as desynchronization or voltage reduction, but also as paradoxical reactivity, K-complexes, or prolonged bursts of delta waves [36].

The EEG in MCS shows diffuse slowing brain activity, mainly of the theta band, and in most cases responsive to external stimuli. However, there are insufficient data as well as the typical pattern of MCS concerns.

EPs may be used to test the integrity of brainstem and forebrain pathways in consciousness disorders.

Although they do not provide reliable information on the location of a lesion in the brainstem, both auditory and somatosensory-evoked potentials, and cortical event-related potentials, can provide information on the prognosis of patients with disorders of consciousness [37].

EPs are affected to varying degrees in VS patients. Although brainstem auditory evoked potentials (BAEPs) can be normal in VS patients [38], the BAEP waveforms can also be attenuated, delayed or absent, depending on the location and degree of brainstem injury [39].

Similarly, somatosensory evoked potentials (SEPs) have been found to be abnormal in PVS patients with delay and attenuation, or even absence of the N20 cortical response to median nerve stimulation [40]. Short-latency evoked potentials have high negative predictive value and bilateral absence of cortical responses of BAEP or SEP and it is regarded as a reliable prognosticator of poor functional outcome. However, their presence does not necessarily indicate a good outcome.

Event-related potentials (ERPs) are more useful than EEG in the differentiation between VS and MCS. Recent data show that the P300 is not a reliable marker of awareness but rather signs automatic processing, as it could be recorded in well documented VS patients who never recovered.

In a recent work, the authors focused on the prediction of consciousness recovery in patients with post-traumatic VS. They used a classical two-stimulus oddball task to elicit the P300 using the patient's own name as deviant and a pure tone as standard stimulus (''subject's own name" paradigm). There is evidence that the amplitude of the P300 wave increases when more salient stimuli are used, such as the own first name instead of visual or auditory deviants. The authors found that P300 is a strong predictor of future recovery of consciousness in VS. This finding is in line with several studies that have confirmed the utility of P300 evoked by deviant tones to predict awakening and favorable outcome from coma and VS [41].

Clinical neurophysiology procedures are useful as easily performed, non invasive and repeatable at the bed side. These methods provide irreplaceable data about the degree of neuronal dysfunctions and their evolution, and gave also information to outcome evaluation.

3. Is the Assessment of Olfactory Function in Consciousness Disorders a Marker of Recovery?

The Olfactory system is very complex. Olfactory perception starts at the level of the olfactory epithelium located in the roof of the nasal cavity. Olfactory receptor neurons (ORNs) are embedded within the nasal mucosa and send their axons through the cribriform plate towards the olfactory bulb. ORNs carry olfactory receptors (ORs) which are key element to olfactory information processing. In the olfactory bulb ORNs axons synapse with second order neurons, called mitral cells. The connection between the olfactory epithelium and the olfactory bulb is characterized by a convergence of axons of ORNs, and all ORNs carrying the same OR converge in the same site within the bulb, called "glomerulus".

The brain structures involved in odor processing mainly consist of the primary olfactory cortex, which comprises the anterior olfactory nucleus, tenia tecta, olfactory tubercole, piriform cortex (PC), anterior cortical amygdaloid nucleus, periamygdaloid and entorhinal cortices. The PC is connected to thalamus, hypothalamus, and orbitofrontal cortex (OFC). The nuclei of the thalamus have further connections towards the OFC and the insular cortex. From the enthorhinal cortex fibres lead to the hippocampus [42-44].

Many factors and pathological conditions may influence normal olfactory function. Recent epidemiological research indicates that at least 5% of the population exhibits a significant loss of olfactory function rendering them functionally anosmic [45-46]. This seems to be largely due to aging. Indeed olfactory decline in healthy subjects strongly correlates with age: between 65 and 80 years about half of the population has experience of a significant decrement in the olfactory ability. Olfactory problems become increasingly significant in the neurological field. Indeed, they occur in a wide range of neurological disorders, such as Alzheimer's and Parkinson's disease, epilepsy, migraine, schizophrenia, olfactory nerve or central olfactory areas injury [47].

To date, nothing is known about the olfactory function of patients with consciousness disorders and its correlation with the clinical severity and outcome of these subjects.

However the assessment of olfactory function in patients with consciousness disorders might provide to the scientific community important information about residual functions of comatose, VS and MCS; in addition, it might support the clinicians in the diagnosis, the assessment of clinical severity and the prediction of patient's outcome.

Olfactory function can be evaluated through the use of specific instrumental approaches, including psychophysical and electrophysiological methods and neuroimaging techniques.

For the clinical assessment of human olfaction, numerous validated psychophysical tests exist. The best-validated olfactory tests include the University of Pennsylvania Smell Identification Test, the Connecticut Chemosensory Clinical Research Center Test and the Sniffin' Sticks Test [48-50]. However, whit these tests the patient cooperation is necessary.

A useful addition for the olfactory function assessment is represented by olfactory event-related potentials (OERPs). OERPs are the result of the sequential activation of numerous brain areas, starting from amygdala and regions of medial temporal lobe, followed by the mid-OFC and insular cortex, along with regions of the temporal lobe.

Unlike the auditory, visual or somatosensory modalities, exploration of the olfactory system using human electrophysiological methods such as event-related potentials has received little attention from the scientific or clinical community. The main reason for that is the lack of adequate methods to produce a selective and controlled stimulation of the olfactory system [51]. Based on the principles of air-dilution olfactometry, Kobal and Platting introduced, in the late 1970s, a chemosensory stimulation with stimuli having a rectangular shape with rapid onset, precisely controlled in terms of timing, duration and intensity and not simultaneously activating other sensory systems [52]. The Olfactometer is a complex instrument for creation of well defined, reproducible smell or pain stimuli in the nose without tactile or thermal stimulation. The IRCCS Centro Neurolesi "Bonino-Pulejo" in Messina has available the Olfactometer, the first and the unique in Italy (Figure 1) and compatible with MR scanner 3 Tesla.

OERPs can be obtained independently of the patient's response bias, allowing the investigation of subjects with difficulty/impossibility to respond properly. Especially in the severe neurological conditions in which it is impossible to have patient's cooperation, OERPs recording could become a useful technique to obtain new information about condition and progression of the clinical picture.

Also functional Magnetic Resonance Imaging (fMRI) is a useful tool for the study of functional neuroanatomy of the human olfactory system.

Several fMRI studies using different delivery systems to administer the odors to the subjects, provided the insights into the brain structures involved in olfactory process. In accordance with the expectations from anatomical data, neuroimaging studies on olfaction reported activation in OFC, PC, amygdala, and entorhinal or parahippocampal gyrus. Other areas have also been consistently found to be activated in response to olfactory stimulation with neuroimaging techniques, including the insula and the anterior cingulated gyrus [53-57]. In addition, fMRI is suitable for studying olfactory information processing. OFC is an area related to tasks involving semantic association or encoding. This area has been reported to be significantly activated during hedonicity judgments for the odor stimuli tasks [57-58]. In particular, Katata et al. [58] reported that left lateral/middle OFC and right lateral OFC were more often activated in the subjects who perceived the odor stimulation as unpleasant.

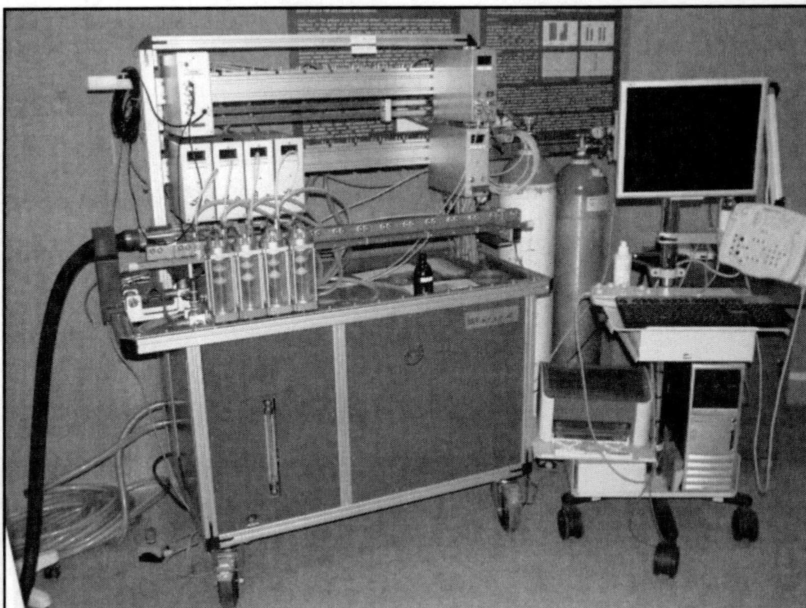

Figure 1. The olfactometer and its accessories available to the IRCCS Centro Neurolesi "Bonino-Pulejo" in Messina.

In addition, this study showed that the right anterior cingulated girus, which is thought to be related to working memory during odor discrimination, was activated more often in subjects who perceived the odor as pleasant. The authors suggested that this area may be involved in the processing of pleasant emotions. Also the insular cortex responds to the hedonics of odors and is involved in the processing of the emotional aspects of odors. In the study of Royet et al. [57] unpleasant odors produced stronger activity than did pleasant odors in the left ventral insula in right-handers subjects and the right ventral insula in left-handers, suggesting lateralized processing of emotional odors as a function of handedness.

Therefore, cerebral imaging findings have shown that olfactory function involves a complex and extensive olfactory neural network. In particular, basic perceptual processes are lateralized between the hemispheres not only at the level of the OFCs, but also in primary olfactory regions including PC and amygdala. While areas located in the right hemisphere such as the OFC and PC are more involved in memory and familiarity ratings, those located in the left hemisphere, such as the OFC, insula, PC, amygdala, temporal lobe and superior frontal cortex, participate more in the emotional response to odours. These different structures would, however, be involved at different levels of emotional olfactory processing. Whereas the piriform–amygdala region appears to be associated with the evaluation of emotional intensity and therefore more activated with unpleasant than pleasant odours, the caudolateral OFC appears to mediate a conscious assessment of these odours. The role of the superior frontal cortex would be to control one's own emotional state in the making of personally relevant decisions [59].

In the context of consciousness disorders studies, fMRI can be used to investigate the brain areas activated by olfactory task. Imaging studies of olfaction require a suitable method for presentation to the subject of odorous stimuli. A MR scanner compatible Olfactometer generates an olfactory stimulus distinguishing between signal (odor condition) and no-signal (control condition without odor) in block-design fMRI studies. The instrument allows an odor generation well standardized with good replication in stimulation exams. FMRI studies of the olfactory system in patients with consciousness disorders might be a useful instrumental approach in order to investigate specific brain areas activated by olfactory stimuli, in particular those implicated in the generation of OERPs and involved in emotional and conscious olfactory processing, with important implications for legal and ethical decision making.

In conclusion, OERPs and fMRI studies in consciousness disorders might help to establish whether the objective measure of olfactory function may constitute useful additional method to neurological and behavioral assessment, improving the differential diagnosis and outcome assessment. Indeed, it is likely that olfactory function is different in VS and MCS and it changes in relation to clinical course of these patients, making its assessment a good predictive marker.

References

[1] Bernat, J. L. Chronic disorders of consciousness. *Lancet,* 2006 367, 2060.

[2] Giacoppo, S; Bramanti, P; Barresi, M; Celi, D; Foti Cuzzola, V; Palella, E; Marino S. Predictive biomarkers *Neurocrit. Care,* 2012 16, 470-477.

[3] Donato, R. S100: a multigenic family of calcium-modulated proteins of the EF-hand type with intracellular and extracellular functional roles. *Int. J. Biochem. Cell Biol.,* 2001 33, 637-668.

[4] Heizmann, C. W; Fritz, G; Schafer, B. W. S100 proteins: structure, functions and pathology. *Front Biosci.,* 2002 7, d1356-1368.

[5] Smit, L. H; Korse, C. M; Hart, A. A; Bonfrer, J. M; Haanen, J. B; Kerst, J. M; Nieweg, O. E; de Gast, G. C. Normal values of serum S-100B predict prolonged survival for stage IV melanoma patients. *Eur. J. Cance.,* 2005 41, 386-392.

[6] Biberthaler, P; Linsenmeier, U; Pfeifer, K. J; Kroetz, M; Mussack, T; Kanz, K. G; Hoecherl, E. F; Jonas, F; Marzi, I; Leucht, P; Jochum, M; Mutschler, W. Serum S-100B concentration provides additional information for the indication of computed tomograghy in patients after minor head injury: a prospective multicenter study. *Shock,* 2006 25, 446-453.

[7] Mussack, T; Biberthaler, P; Kanz, K. G; Wiedemann, E; Gippner-Steppert, C; Mutschler, W; Jochum, M. Serum S-100B and interleukin-8 as predictive markers for comparative neurologic outcome analysis of patients after cardiac arrest and severe traumatic brain injury. *Crit. Care Med.,* 2002 30, 2669-2674.

[8] Ingebrigtsen, T; Romner B. Biochemical serum markers of traumatic brain injury. *J. Trauma,* 2002 52, 798-808.

[9] Raabe, A; Grolms, C; Keller, M; Döhnert, J; Sorge, O; Seifert, V. Correlation of computed tomography findings and serum brain damage markers following severe head injury. *Acta Neurochir.,* 1998 140, 787-791.

[10] Woertgen, C; Rothoerl, R. D; Metz, C; Brawanski, A.Comparison of clinical, radiologic, and serum marker as prognostic factors after severe head injury. *J. Trauma*, 1999 47, 1126-1130.

[11] Herrmann, M; Curio, N; Jost, S. Release of biochemical markers of damage to neuronal and glial brain tissue is associated with short and long term neuropsychological outcome after traumatic brain injury. *J. Neurol. Neurosurg. Psychiatry,* 2001 70, 95-100.

[12] Meynaar, I. A; Oudemans-van Straaten, H. M; van der Wetering, J; Verlooy, P; Slaats, E. H; Bosman, R. J; van der Spoel, J. I; Zandstra, D. F.Serum neuron-specific enolase predicts outcome in post-anoxic coma: a prospective cohort study. *Intensive Care Med.* 2003 29, 189-195.

[13] Vos, P. E; Lamers, K. J; Hendriks, J. C; van Haaren, M; Beems, T; Zimmerman, C; van Geel, W; de Reus, H; Biert, J; Verbeek, M. M. Glial and neuronal proteins in serum predict outcome after severe traumatic brain injury. *Neurology,* 2004 62, 1303-1310.

[14] Berger, R. P; Dulani, T; Adelson, P. D; Leventhal, J. M; Richichi, R; Kochanek, P. M. Identification of inflicted traumatic brain injury in well-appearing infants using serum and cerebrospinal markers: a possible screening tool. *Pediatrics,* 2006 117, 325-332.

[15] Johnsson, P. Markers of cerebral ischemia after cardiac surgery. *J. Cardiothorac. Vasc. Anesth.*, 1996 10, 120-126.

[16] Pelinka, L. E; Hertz, H; Mauritz, W; Harada, N; Jafarmadar, M; Albrecht, M; Redl, H; Bahrami, S. Nonspecific increase of systemic neuron-specific enolase after trauma: clinical and experimental findings. *Shock,* 2005 24, 119-123.

[17] Berger, R. P; Adelson, P. D; Pierce, M. C; Dulani, T; Cassidy, L. D; Kochanek, P. M. Serum neuron-specific enolase, S100B, and myelin basic protein concentrations after inflicted and noninflicted traumatic brain injury in children. *J. Neurosurg.,* 2005 103, 61–68.

[18] Thomas, D. G; Palfreyman, J. W; Ratcliffe, J. G. Serum-myelin-basic-protein assay in diagnosis and prognosis of patients with head injury. *Lancet,* 1978 1, 113-115.

[19] Yamazaki Y, Yada K, Morii S, Kitahara T, Ohwada T. Diagnostic significance of serum neuron-specific enolase and myelin basic protein assay in patients with acute head injury. *Surg. Neurol.,* 1995 Mar;43(3):267-70.

[20] Herrmann, M; Vos, P; Wunderlich, M. T; de Bruijn, C. H; Lamers, K. J. Release of glial tissue-specific proteins after acute stroke: A comparative analysis of serum concentrations of protein S-100B and glial fibrillary acidic protein. *Stroke,* 2000 31, 2670-2677.

[21] Missler, U; Wiesmann, M; Wittmann, G; Magerkurth, O; Hagenström, H. Measurement of glial fibrillary acidic protein in human

blood:analytical method and preliminary clinical results. *Clin. Chem.,* 1999 45, 138–141.

[22] Nylén, K; Ost, M; Csajbok, L. Z; Nilsson, I; Blennow, K; Nellgård, B; Rosengren, L. Increased serum-GFAP in patients with severe traumatic brain injury is related to outcome. *J. Neurol. Sci.,* 2006 240, 85-91.

[23] Campello Yurgel, V; Ikuta, N; Brondani da Rocha, A; Lunge, V. R; Fett Schneider, R; Kazantzi Fonseca, A. S; Grivicich, I; Zanoni, C; Regner, A. Role of plasma DNA as a predictive marker of fatal outcomes following severe head injury in males. *J. Neurotrauma*, 2007 24, 1172-1181.

[24] Lo,Y. M; Rainer, T. H; Chan, L. Y; Hjelm, N. M; Cocks, R. A. Plasma DNA as a prognostic marker in trauma patients. *Clin. Chem.,* 2000 46, 319-323.

[25] Lam, N. Y; Rainer, T. H; Chan, L. Y; Joynt, G. M; Lo, Y. M. Time course of early and late changes in plasma DNA in trauma patients. *Clin. Chem.,* 2003 49, 1286-1291.

[26] McAllister, A. K; Katz, L. C; Lo, D. C. Neurotrophins and synaptic plasticity. *Annu. Rev. Neurosci.,* 1999 22, 295–318.

[27] Tyler, W. J; Alonso, M; Bramham, C. R; Pozzo-Miller, L. D. From acquisition to consolidation: on the role of brain-derived neurotrophic factor signaling in hippocampal-dependent learning. *Learn Mem.,* 2002 9, 224–237.

[28] Riley, C. P; Cope, T. C; Buck, C. R. CNS neurotrophins are biologically active and expressed by multiple cell types. *J. Mol. Histol.,* 2004 35, 771–783.

[29] Saha, R. N; Liu, X; Pahan, K. Up-regulation of BDNF in astrocytes by TNF-alpha: a case for the neuroprotective role of cytokine. *J. Neuroimmune Pharmacol.,* 2006 1, 212-222.

[30] Jackson, P; Thompson, R. J. The demonstration of new human brain-specific proteins by high-resolution two-dimensional polyacrylamide gel electrophoresis. *J.Neurol. Sci.,* 1981 49, 429-438.

[31] Papa, L; Akinyi, L; Liu, M. C; Pineda, J. A; Tepas, J. J 3rd; Oli, M. W; Zheng, W; Robinson, G; Robicsek, S. A; Gabrielli, A; Heaton, S. C; Hannay, H. J; Demery, J. A; Brophy, G. M; Layon, J; Robertson, C. S; Hayes, R. L; Wang, K. K.Ubiquitin Cterminal hydrolase is a novel biomarker in humans for severe traumatic brain injury. *Crit. Care Med.,* 2010 38, 138-144.

[32] Coplin, W. M; Longstreth, W. T Jr; Lam, A. M; Chandler, W. L; Mayberg, T. S; Fine, J. S; Winn, H. R. Cerebrospinal fluid creatine

kinase-BB isoenzyme activity and outcome after subarachnoid hemorrhage. *Arch. Neurol.,* 1999 56, 1348-1352.

[33] Bell, R. D; Rosenberg, R. N; Ting, R; Mukherjee, A; Stone, M. J; Willerson, J. T. Creatine kinase BB isoenzyme levels by radioimmunoassay in patients with neurological disease. *Ann. Neurol.,* 1978 3, 52-59.

[34] Hughes, J. R. Limitations of the EEG in coma and brain death. *Ann. N Y Acad. Sci.,* 1978 315, 121-136.

[35] Babiloni, C; Sarà, M; Vecchio, F; Pistoia, F; Sebastiano, F; Onorati, P; Albertini, G; Pasqualetti, P; Cibelli, G; Buffo, P; Rossini, P. M. Cortical sources of resting-state alpha rhythms are abnormal in persistent vegetative state patients. *Clin. Neurophysiol.,* 2009 120, 719-729.

[36] Guérit, J. M; Amantini, A; Amodio, P; Andersen, K. V; Butler, S; de Weerd, A; Facco, E; Fischer, C; Hantson, P; Jäntti, V; Lamblin, M. D; Litscher, G; Péréon, Y.Consensus on the use of neurophysiological tests in the intensive care unit (ICU): electroencephalogram (EEG), evoked potentials (EP), and electroneuromyography (ENMG). *Neurophysiol. Clin.,* 2009 39, 71-83.

[37] Fischer, C; Luauté, J; Adeline, P; Morlet, D. Predictive value of sensory and cognitive evoked potentials for awakening from coma. *Neurology,* 2004 63, 669-673.

[38] Hansotia, P. L. Persistent vegetative state. Review and report of electrodiagnostic studies in eight cases. *Arch. Neurol.,* 1985 42, 1048–1052.

[39] Li, S; Wei, J; Guo, D. Persistent vegetative state: Clinical and electrophysiologic observations of 5 cases. *Chin. Med. Sci. J.,* 1993 8, 101–106.

[40] Isono, M; Wakabayashi, Y; Fujiki, M. M; Kamida, T; Kobayashi, H. Sleep cycle in patients in a state of permanent unconsciousness. *Brain Inj.,* 2002 16, 705–712.

[41] Cavinato, M; Freo, U; Ori, C; Zorzi, M; Tonin, P; Piccione, F; Merico, A. Post-acute P300 predicts recovery of consciousness from traumatic vegetative state. *Brain Inj.,* 2009 23, 973-980.

[42] Gottfried, JA; Winston, JS; Dolan, RJ. Dissociable codes of odor quality and odorant structure in human piriform cortex. *Neuron,* 2006 49, 467-479.

[43] Zelano, C; Montag, J; Johnson, B; Khan, R; Sobel, N. Dissociated representations of irritation and valence in human primary olfactory cortex. *J. Neurophysiol.,* 2007 97, 1969-1976.

[44] Bensafi, M; Sobel, N; Khan, R. M. Hedonic-specific activity in piriform cortex during odor imagery mimics that during odor perception. *J. Neurophysiol.*, 2007 98, 3254-3262.

[45] Landis, B. N; Konnerth, C. G; Hummel, T. A study on the frequency of olfactory dysfunction. *Laryngoscope*, 2004 114, 1764-1769.

[46] Brämerson, A; Johansson, L; Ek, L; Nordin, S; Bende, M. Prevalence of olfactory dysfunction: the skövde population-based study. *Laryngoscope,* 2004; 114, 733-737.

[47] Doty, RL. The olfactory system and its disorders. *Semin. Neurol.*, 2009 29:74-81.

[48] Cain, W. S; Gent, J. F; Goodspeed, R. B; Leonard, G. Evaluation of olfactory dysfunction in the Connecticut Chemosensory Clinical Research Center. *Laryngoscope,* 1988 98, 83-88.

[49] Doty, R. L; Shaman, P; Kimmelman, C. P; Dann, M. S. University of Pennsylvania Smell Identification Test: a rapid quantitative olfactory function test for the clinic. *Laryngoscope,* 1984 94, 176-178.

[50] Hummel, T; Sekinger, B; Wolf, S. R; Pauli, E; Kobal, G. 'Sniffin' sticks': olfactory performance assessed by the combined testing of odor identification, odor discrimination and olfactory threshold. *Chem. Senses.*, 1997 22, 39-52.

[51] Rombaux, P; Mouraux, A; Bertrand, B; Guerit, J. M; Hummel, T. Assessment of olfactory and trigeminal function using chemosensory event-related potential *Neurophysiol. Clin.,* 2006 36, 53-62.

[52] Kobal, G; Plattig, K. H. Objective olfactometry: methodological annotations for recording olfactory EEG-responses from the awake human. *EEG EMG.* 1978 9, 135-145.

[53] Koizuka, I; Yano, H; Nagahara, M; Mochizuki, R; Seo, R; Shimada, K; Kubo, T; Nogawa, T. Functional imaging of the human olfactory cortex by magnatic resonance imaging. ORL. *Otorhinolaryngol. Relat. Spec.,* 1994 56, 273-275.

[54] Fulbright, R. K; Skudlarski, P; Lacadie, C. M; Warrenburg, S; Bowers, A. A; Gore, J. C; Wexler, B. E. Functional M. R imaging of regional brain responses to pleasant and unpleasant odors. *AJNR Am. J. Neuroradiol.,* 1998 19, 1721–1726.

[55] Sobel N, Prabbakaran V, Zhao Z, et al. Time course of odorant-induced activation in the human primary olfactory cortex. *J. Neurophysiol.,* 2000; 83(1): 537-51.

[56] Zald, D. H; Pardo, J. V. Functional neuroimaging of the olfactory system in humans. *Int. J. Psychophysiol.,* 2000 36, 165-168.

[57] Royet, J. P; Plailly, J; Delon-Martin, C; Kareken, D. A; Segebarth, C. fMRI of emotional responses to odors: influence of hedonic valence and judgment, handedness, and gender. *NeuroImage,* 2003 20, 713–728.

[58] Katata, K; Sakai, N; Doi, K; Kawamitsu, H; Fujii, M; Sugimura, K; Nibu, K. Functional MRI of region brain responses to "pleasant" and "unpleasant" odors. *Acta Oto-Laryngologica,* 2009 129, 85-90.

[59] Royet, JP; Plailly, J. Lateralization of olfactory processes. *Chemical. Senses.,* 2004; 29, 731–745.

In: Chronic Disorders of Consciousness ISBN: 978-1-62618-940-9
Editors: R. S. Calabrò and P. Bramanti

Chapter 6

The Pharmacotherapy of "Awakening"

Rocco Salvatore Calabrò, Patrizia Pollicino and Placido Bramanti
IRCCS Centro Neurolesi, Messina, Italy

Abstract

Disorders of consciousness (DOC) include coma, vegetative state (VS) and minimally conscious state (MCS). As revealed by functional imaging studies, clinical recovery is often associated with a functional restoration of cortico-thalamo-cortical connections. Depending on the amount of network restored, patients may regain full consciousness or remain in a MCS. Molecular and neural mediators may indirectly contribute to the above restoration processes owing to their role in the phenomenon of neural synaptic plasticity. Therefore, there is growing interest in the possible effects of drugs that act at the level of the CNS in promoting emergence from DOC. Pharmacological treatment has been attempted with various drugs, including levodopa, amphetamines, tricyclic antidepressants, amantadine, bromocriptine, anticonvulsants, and others. Levodopa and amantadine might be useful when parkinsonian features are present, but they are more likely to benefit patients in minimally conscious state than those in vegetative state. The present chapter is aimed at investigating the complex pharmacological management of DOC.

Introduction

Progress in emergency treatment is responsible for the recovery of many patients daily who would have died otherwise. However, as a result, the number of patients in a VS and in a MCS has increased. Persistent VS is a pathologic condition of complete unawareness of the self and the environment, accompanied by sleep-wake cycles, with preservation of brainstem autonomic and hypothalamic function as a result of a disconnection of long-range corticocortical (between latero-frontal and midline-posterior areas) and cortico-thalamo-cortical (between non-specific thalamic nuclei and lateral and medial frontal cortices) pathways, whereas MCS is characterized by inconsistent, but clearly discernible, behavioral evidence of consciousness, with a partial preservation of connectivity among cortico-cortical and cortico-thalamocortical networks. The causes of DOC are several and include traumatic brain injury, hypoxic damage, cerebrovascular ischemic or hemorrhagic injury, infection of the central nervous system, toxins, poisoning, degenerative diseases, tumors as well as congenital or developmental disorders. When an acquired brain injury results in DOC, coma is the initial clinical manifestation: usually, it evolves within 2-4 weeks into either a conscious state or VS or brain death. Clinical recovery depends on the residual synergy of widely interconnected cortical-subcortical networks, in accordance with the concept of functional connectivity. From a biochemical perspective, the regaining of a balance between molecular and neural mediators normally acting at the level of the CNS may lay the groundwork for the recovery of synaptic homeostasis that, in turn, may produce a degree of functional connectivity that is sufficient for clinical improvement.

Medical Treatments

After discharge from the intensive care unit, patients in VS require the same level of medical and nursing care as comatose patients, including, positioning, range of motion exercises, pulmonary toilet, skin care, bowel and bladder care, and optimum nutrition. All patients require gastrostomy for tube feedings and many of them have a tracheostomy for airway protection, but no longer require mechanical ventilation. Spasticity is often severe and requires aggressive treatment, including intrathecal administration of baclofen and/or use of botulin toxins. Attempts to improve the responsiveness of the patient in

VS with sensory, electrical, and pharmacological stimulation have been disappointing, but most studies represent class III evidence, making firm conclusions hard to draw [1].

1. CNS Stimulants

Drugs with a pharmacological profile compatible with that of a CNS stimulant and that have been studied in DOC include antidepressants, dopaminergic agents and drugs used to treat attention deficit hyperactivity disorders, i.e. methylphenidate and pure stimulants.

a. Antidepressants

Previous studies have shown that both tricyclic antidepressants (TCAs), i.e. protriptyline, amitriptyline and desipramine, and selective serotonin reuptake inhibitors (SSRIs), such as sertraline can improve arousal or initiation in vegetative patients following severe brain injury [2-4]. The majority of the TCAs act primarily as serotonin-norepinephrine reuptake inhibitors (SNRIs) by blocking the serotonin transporter (SERT) and the norepinephrine transporter (NET), respectively, resulting in an elevation of the synaptic concentrations of these neurotransmitters, and therefore an enhancement of neurotransmission. Notably, the TCAs have negligible affinity for the dopamine transporter (DAT), and therefore have no efficacy as dopamine reuptake inhibitors (DRIs). Both serotonin and norepinephrine have been highly implicated in depression and anxiety, and it has been shown that facilitation of their activity has beneficial effects on these mental disorders. In addition to their reuptake inhibition, many TCAs also have high affinity as antagonists at the 5-HT_2, 5-HT_6, 5-HT_7, α_1-adrenergic, and NMDA receptors, and as agonists at the sigma receptors, some of which may contribute to their therapeutic efficacy, as well as their side effects. The TCAs also have varying but typically high affinity for antagonising the H_1 and H_2 histamine receptors, as well as the muscarinic acetylcholine receptors. As a result, they also act as potent antihistamines and anticholinergics. These properties are generally undesirable in antidepressants, and are likely to contribute to their large side effect profiles. Most, if not all, of the TCAs also potently inhibit sodium channels and L-type calcium channels, thus acting as sodium and calcium channel blockers and calcium channel blockers, respectively.

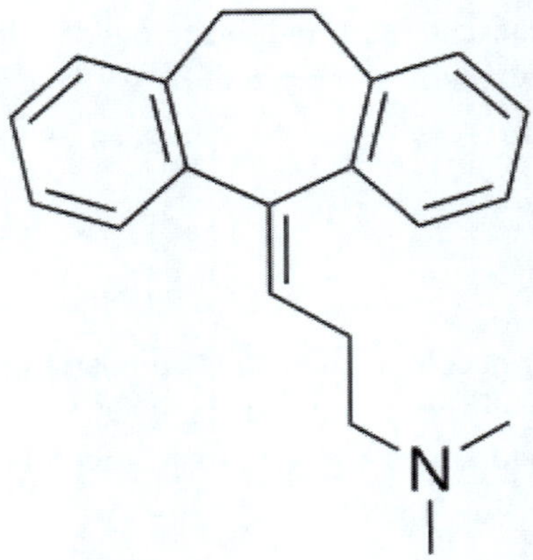

Figure 1. Chemical structure of the tricyclic antidepressant amitriptyline.

Interestingly, the sodium channel blockage is responsible for the high mortality rate upon overdose seen with the TCAs via cardiotoxicity. Amitriptyline (Figure 1) and desipramine showed cognitive enhancing effects in isolated cases [2]. Similarly, protriptlyline has been described as an efficient activating/ stimulant medication in eight individual case reports [5]. On the other hand, a recent prospective, randomized, placebo-controlled trial failed to demonstrate a sertraline-induced improvement in 11 severely brain-injured patients [3]. Sertraline belongs to the SSRIs, a class of compounds typically used as antidepressants in the treatment of depression, anxiety disorders, and some personality disorders. SSRIs are believed to increase the extracellular level of the neurotransmitter serotonin by inhibiting its reuptake into the presynaptic cell, increasing the level of serotonin in the synaptic cleft available to bind to the postsynaptic receptor. They have varying degrees of selectivity for the other monoamine transporters, with pure SSRIs (such as citalopram and escitalopram) having only weak affinity for the noradrenaline and dopamine transporter. After the use of an SSRI, since there is more serotonin available, the response is to decrease the number of postsynaptic receptors over time and in the long run, thus modifying the serotonin/receptor ratio. This down-regulation of 5-HT2_A occurs when the antidepressant effects of SSRIs become apparent. Notably, deceased suicidal and otherwise depressed patients have had more 5-HT2A receptors than normal patients, suggesting that 5-HT2_A overactivity is involved in the pathogenesis of depression. However, within the DOC context, it is worth drawing attention to the brain circuitry that regulates the transition between sleep and wakefulness. In this regard, wakefulness is regulated by the ascending arousal system. It is intuitive that drugs that can interact with the circuitry might alter the transition between sleep and wakefulness, as may be hypothesized in the case of antidepressants influencing noradrenergic and serotonergic systems and subsequently promoting the ascending arousal pathways. Sleep-wake cycles, although preserved in patients in persistent VS and MCS, are commonly irregular and may negatively affect brain plasticity phenomena leading to functional recovery [6-7]. Future studies will evaluate whether drugs that are capable of favouring the stabilization and coherence of the circadian rhythm in patients in persistent VS and MCS may indirectly orientate neural repair processes.

b. Dopaminoagonists

Brain injuries often result in diffuse axonal injury that may be associated with lower dopamine turnover and a decrease in CSF dopamine level [8] The dopaminergic pathways mainly involved after a traumatic injury include the nigrostriatal one that plays important role in initiation and speed of movements; the mesolimbic pathway that is responsible for emotion, motivation, learning and memory; and the mesocortical one that is connected with frontal lobes and regulates motivation, planning and attention. As dopamine is the key neurotransmitter in the regulation of behaviour, mood, language, cognition, arousal and motor control, the inhibition of dopaminergic stimulation may be associated with impaired neurological functions in patients with DOC, thus providing the rationale for the use of dopaminergic agents to sustain their recovery. *L-DOPA* (L-3,4-dihydroxyphenylalanine) is a chemical that is made and used as part of the normal biology of some animals and plants (Figure 2). Some animals including humans make it via biosynthesis from the amino acid L-tyrosine. L-DOPA is the precursor to the neurotransmitters dopamine, norepinephrine (noradrenaline), and epinephrine (adrenaline) collectively known as catecholamines. Moreover, L-DOPA can be manufactured and in its pure form is sold as a psychoactive drug with the INN levodopa, used in the clinical treatment of Parkinson's disease and dopamine-responsive dystonia. However, previous studies have demonstrated the efficacy of levodopa in treating subgroups of patientswith DOC: levodopa has been reported to promote a transition toward a full consciousness condition or a MCS [9-12]. Among patients with VS and MCS, those more likely to respond to levodopa show concomitant symptoms of parkinsonism (particularly unilateral predominant rigidity/ spasticity and akinesia) and neuroradiological findings of high-intensity lesions in the dopaminergic pathway [10].

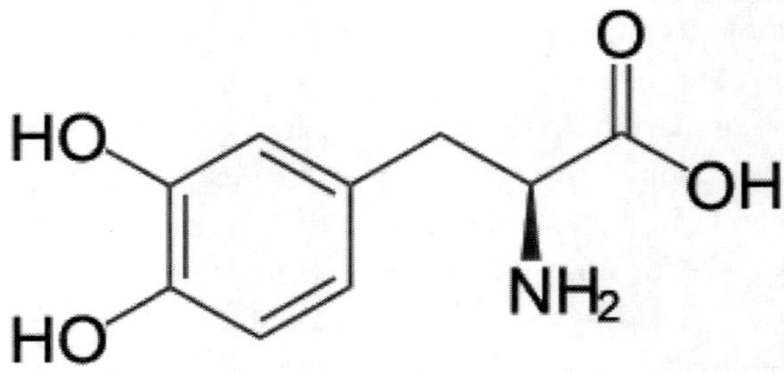

Figure 2. Levodopa.

Hence, patients in persistent VS or MCS in whom the dopaminergic pathways may have been selectively damaged seem to be particularly responsive to levodopa. *Apomorphine* is a well recognized dopaminergic agonist, the use of which is limited by the fact that it produces nausea and emesis. A single case report on the effect of apomorphine in a patient in a MCS further supports the concept that a dopaminergic deficit may underlie consciousness impairment after traumatic brain injury [13]. In support of this finding, a prospective case series recently showed that continuous subcutaneous administration of apomorphine resulted in a consciousness improvement in most treated patients [14]. *Pramipexole*, which is a selective agonist for dopamine D_2 receptors and is mainly used for treating Parkinson's disease and restless legs syndrome, has been recognized as being useful in promoting the recovery of children with persistent VS [15]. Similarly, *bromocriptine*, which is an agonist of post-synaptic D_2 receptors, was shown to improve alertness, arousal and cognitive awareness in five patients with persistent VS after traumatic brain injury [16]. *Amantadine* acts at both the pre-synaptic level, by inhibiting dopamine re-uptake, and the post-synaptic level, by increasing the density and altering the conformation of dopamine receptors (Figure 3). Moreover, amantadine is an NMDA receptor antagonist. Although some studies [8, 17, 18] suggested the usefulness of amantadine in minimally conscious and vegetative patients, they have some limitations that prevent the extrapolation of results to a larger population of unconscious patients. However, a recent study by Giacino et al. [19] accelerated the pace of functional recovery during active treatment in patients with post-traumatic DOC. The authors enrolled 184 patients who were in VS or MCS (4 to 16 weeks after traumatic brain injury), and who were receiving inpatient rehabilitation; patients were randomly assigned to receive amantadine or placebo for 4 weeks and were followed for 2 weeks after the treatment was discontinued.

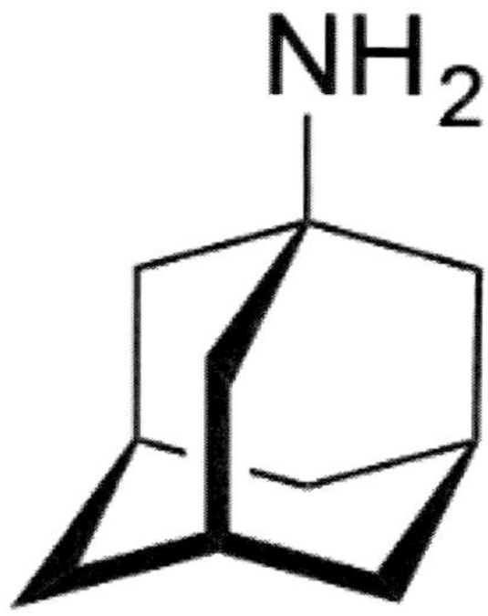

Figure 3. amantadine chemical structure.

During the 4-week treatment period, recovery was significantly faster in the amantadine group than in the placebo group, as measured by the Disability Rating Scale score.

c. Others

Methylphenidate, a nonspecific pre-synaptic catecholamine mimetic, has been reported to facilitate arousal in patients with brain injury [20]. Indeed, some authors have shown a favourable effect of the drug in semi-comatose and comatose patients who show a low level of consciousness. These observations suggest a putative effect of the drug on different domains, such as awareness or wakefulness or both, that are beyond the cognitive one. On the other hand, amfetamine, a psychostimulant drug that, by increasing the levels of dopamine in the synaptic cleft, results in enhanced wakefulness and decreased fatigue and appetite, seems to improve both arousal and motor recovery in post-stroke and traumatic brain injury patients [20].

2. CNS Depressants

The finding that sedative drugs can improve the state of consciousness may appear paradoxical; however, several cases of improvement, with a recovery of consciousness following shortly after the start of such treatments, have been recently reported. CNS depressants that have been associated with recovery from DOC include zolpidem (a sedative-hypnotic), baclofen (a muscle relaxant) and lamotrigine (an antiepileptic drug).

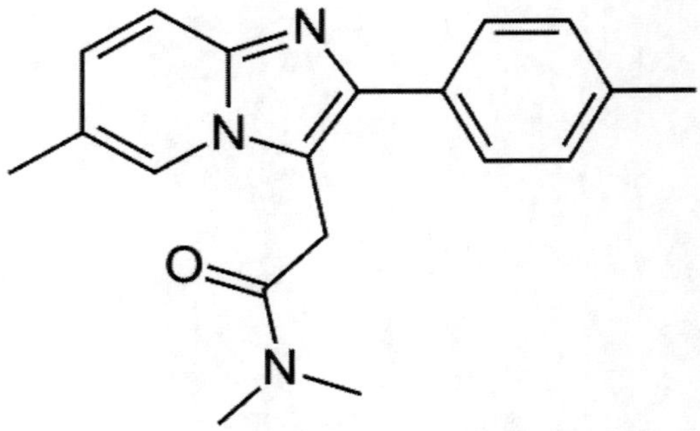

Figure 4. zolpidem chemical structure.

a. Zolpidem

Zolpidem is a non-benzodiazepine sedative/hypnotic belonging to the imidazopirine class, which is chemically distinct from other hypnotics (Figure 4). Due to its selective binding, Zolpidem has very weak anxiolytic, myorelaxant and anticonvulsant properties (but only at higher dosages), but very strong hypnotic properties. Zolpidem binds with high affinity and acts as a full agonist at the α_1-containing $GABA_A$ receptors, about 10-fold lower affinity for those containing the α_2- and α_3- $GABA_A$ receptor subunits, and with no appreciable affinity for α_5 subunit-containing receptors. While the ω_1 type $GABA_A$ receptors are the α_1-containing $GABA_A$ receptors, the ω_2 $GABA_A$ receptors are the α_2-, α_3-, α_4-, α_5-, and α_6-containing $GABA_A$ receptors. The ω_1 $GABA_A$ receptors are found primarily in the brain, whereas the ω_2 ones are found primarily in the spine. Thus, zolpidem has a preferential binding for the $GABA_A$-benzodiazepine receptor complex in the brain but a low affinity for the $GABA_A$-benzodiazepine receptor complex in the spine.

Previously reported data show cases of arousal from persistent VS and functional improvement in MCS after zolpidem administration for sleep-wake cycle disturbances or restlessness. [21-23] For instance, three young patients in persistent VS following traumatic brain injury were described as awakening and remaining awake during the period that the drug would have had a pharmacological effect (with maximum arousal seen 1 hour after the start of the treatment) [22]. Four hours after administration, when the drug action would have subsided, the patients returned to their VS. Interestingly, all of these patients had received daily treatment with zolpidem for several years and on each day they had remained awake during the period that the drug would have had an effect. Similar clinical improvements without remarkable side effects were reproduced in another 3 studies after daily administration of 5-10 mg of zolpidem in a total of 3 patients suffering from postanoxic DOC (disease duration: 8–24 months) [23-25]. Of these studies, the work by Brefel-Courbon et al. [24] was of particular interest as it evaluated zolpidem in a randomized, double-blind, placebo- controlled, single-patient trial: zolpidem was administered to a 48-year-old woman who had been suffering from akinetic mutism for over 2 years, with a marked improvement in motor performance and neuropsychological tests, as well as in the cerebral metabolism of the postrolandic territories and frontal cortex, and of the anterior cingulate and orbitofrontal cortices.

A putative mechanism that explains zolpidem related arousal has been proposed. After brain injury or cerebral anoxic damage there is a change in the levels of glutamate and GABA neurotransmission, which may either be a consequence of the damage or a form of biological adaptation [21, 22]. In particular, a concomitant enhancement in the release of glutamate that results in excitotoxicity and subsequent apoptosis, and an overflow of GABA and long-term supersensitivity of GABAA receptors, have been observed. In the latter case, the inhibitory neurotransmitter binds to its ion channel receptors, which results in reduced cell metabolism and blood flow in brain areas adjacent to the damaged area (state of cell dormancy to contrast hypoxia). By binding to modified $GABA_A$ receptors of the neurodormant cells, zolpidem may cause reversal of the abnormal state and associated metabolic inhibition. Contrary to the previously described positive effects of zolpidem, Whyte and Myers [26] have recently reported that in their double-blind, placebo-controlled, crossover design study on zolpidem (10 mg/day) and placebo, only 1 patient exhibited a clinically significant improvement, out of 14 patients who had been in either VS or MCS due to trauma, anoxia or stroke. Notably, that single patient reproducibly improved from VS to MCS; in the remaining 13 patients, not even a subclinical response was observed, indicating a bimodal type of response rather than a scaled one.

b. Baclofen

The precise mechanism of action of baclofen (Figure 5) has not been fully determined: it exerts its effects as an agonist at presynaptic GABA-B (bicuculline-insensitive) receptors and acts mainly at the spinal cord level inhibiting the transmission of both monosynaptic and polysynaptic reflexes, possibly by hyperpolarization of primary afferent fiber terminals, resulting in antagonism of the release of putative excitatory transmitters (i.e., glutamic and aspartic acids). Postsynaptic effects of baclofen have been demonstrated at concentrations above the therapeutic range. In addition, actions at the supraspinal sites may be involved. Baclofen has general CNS-depressant actions. Animal studies demonstrated that baclofen also may have antinociceptive effects, possibly by acting at spinal and supraspinal sites by decreasing the responsiveness of nociceptive neurons, thereby resulting in analgesia.

Sporadic cases of dramatic recovery from persistent VS after intrathecal baclofen (ITB) administration for spasticity have been reported, which suggests that the drug may play an important role in both arousal and improved awareness [27-30]

The reported improvements ranged from a mere increase in alertness to a full recovery of consciousness, and from transient (i.e. an improvement followed by a return to the previous level of unconsciousness) to sustained. Two mechanisms may underlie the effect of ITB in improving the state of consciousness: (i) a modulation of spinal cord segmental activities that gives rise to centripetal outputs toward the cortex; and (ii) a modulation of sleep-wake cycles that, although present in patients with persistent VS, may be dysregulated and interfere with basic alertness and the recovery of consciousness.

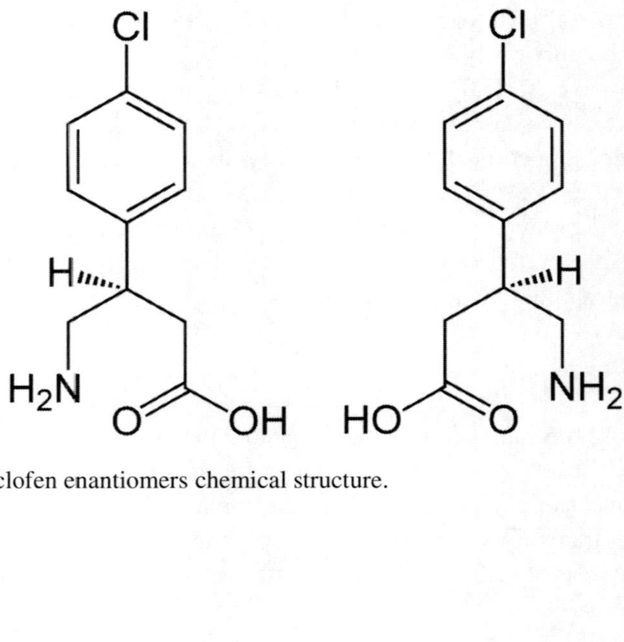

Figure 5. Baclofen enantiomers chemical structure.

Other Treatment Options

Non pharmacological treatment options [31] include:

- *Median nerve stimulation* has been tested only in a patient [32], who had been in postanoxic VS for 6 months, received right median nerve stimulation, with positive responses including better posture, visual pursuit and fixation, swallowing and phonation, within the first 3 months of treatment, and with no major complications.
- In a 6-month, double-blind, alternating crossover study, Schiff et al. [33] showed that bilateral *deep brain stimulation* (DBS) of the anterior intralaminar nuclei of the thalamus, can induce behavioral responsiveness in a patient who had remained in MCS for 6 years before the intervention due to a traumatic brain injury.
- Recently, Canavero et al. [34] have demonstrated that unilateral, bifocal, *extradural cortical stimulation* (ECS) could produce a therapeutic result in a patient who had been in posttraumatic VS for a period of 20 months. The investigators chose to stimulate the sulcus between the left parietal gyri P1 and P2 and the middle frontal sulcus (F2), including the dorsolateral prefrontal cortex (Brodmann's areas 8 and 46).
- The effects of *spinal cord stimulation* (SCS) on 214 patients who had been in VS due to trauma, anoxia or stroke were studied in a prospective observational study for 20 consecutive years (1986–2005) [35]. The SCS device was implanted into the epidural space between C2 and C4, stimulating the spinal cord in a cyclic mode of 15 min on and 15 min off only during daytime, without reaching the motor threshold. Interestingly, a clinical response was recorded in 109 of the 201 patients (54%; 13 were lost to follow-up), and it was correlated with age below 35, traumatic brain injury and regional cerebral blood flow over 20 ml/100 g/min.

Conclusion

DOC should be interpreted along a clinical *continuum r*anging from coma to consciousness recovery. Unfortunately, to date, there is no accepted pharmacological therapy for these clinical conditions. Indeed, the treatments proposed for DOC have not yet gained the level of evidence-based treatment; moreover, the studies to date have led to inconclusiveness. On the other hand, recent serendipitous observations suggest that specific agents may act on neurotransmitter-neurofunctional networks that might contribute to consciousness recovery. However, this needs to be established by further clinical studies of sound methodology, e.g. prospective cohort studies or double-blind placebo-controlled studies. The mechanisms of action and limitations of the suggested treatments should be further investigated.

References

[1] Pistoia F, Mura E, Govoni S, Fini M, Sarà M. Awakenings and awareness recovery in disorders of consciousness: is there a role for drugs? *CNS Drugs*. 2010; 24(8):625-38.

[2] Reinhard DL, Whyte J, Sandel ME. Improved arousal and initiation following tricyclic antidepressant use in severe brain injury. *Arch. Phys. Med. Rehabil.* 1996; 77: 80-3.

[3] Meythaler JM, Depalma L, Devivo MJ, et al. Sertraline to improve arousal and alertness in severe traumatic brain injury secondary to motor vehicle crashes. *Brain Inj.* 2001; 15: 321-31.

[4] Jackson RD, Corrigan JD, Arnett JA. Amitriptyline for agitation in head injury. *Arch. Phys. Med. Rehabi.* l 1985; 66: 180-1.

[5] Wroblewski B, Glenn MB, Cornblatt R, et al. Protriptyline as an alternative stimulant medication in patients with brain injury: a series of case reports. *Brain Inj.* 1993; 7: 353-62.

[6] D'Aleo G, Saltuari L, Gerstenbrand F, et al. Sleep in the last remission stages of vegetative state of traumatic nature. *Funct Neurol* 1994; 9: 189-92.

[7] Isono M, Wakabayashi Y, Fujiki MM, et al. Sleep cycle in patients in a state of permanent unconsciousness. *Brain Inj.* 2002; 16: 705-12.

[8] Meythaler JM, Brunner RC, Johnson A, et al. Amantadine to improve neurorecovery in traumatic brain injuryassociated diffuse axonal injury: a pilot double-blind randomized trial. *J. Head Trauma Rehabil.* 2002; 17: 300-13.

[9] Haig AJ, Ruess JM. Recovery from vegetative state of six months' duration associated with Sinemet (levodopa/ carbidopa). *Arch. Phys. Med. Rehabil.* 1990; 71: 1081-3.

[10] Matsuda W, Matsumura A, Komatsu Y, et al. Awakenings from persistent vegetative state: report of three cases with parkinsonism and brain stem lesions on MRI. *J. Neurol. Neurosurg. Psychiatry* 2003; 74: 1571-3.

[11] Krimchansky BZ, Keren O, Sazbon L, et al. Differential time and related appearance of signs, indicating improvement in the state of consciousness in vegetative state traumatic brain injury (VS-TBI) patients after initiation of dopamine treatment. *Brain Inj.* 2004; 18: 1099-105.

[12] Matsuda W, Komatsu Y, Yanaka K, et al. Levodopa treatment for patients in persistent vegetative or minimally conscious states. *Neuropsychol. Rehabil.* 2005; 15: 414-27.

[13] Fridman EA, Krimchansky BZ, Bonetto M, et al. Continuous subcutaneous apomorphine for severe disorders of consciousness after traumatic brain injury. *Brain Inj.* 2010; 24: 636-41.

[14] Fridman EA, Calvar J, Bonetto M, et al. Fast awakening from minimally conscious state with apomorphine. *Brain Inj.* 2009; 23: 172-7.

[15] Patrick PD, Buck ML, Conaway MR, et al. The use of dopamine enhancing medications with children in low response states following brain injury. *Brain Inj.* 2003; 17: 497-506.

[16] Passler MA, Riggs RV. Positive outcomes in traumatic brain injury-vegetative state: patients treated with bromocriptine. *Arch. Phys. Med. Rehabil.* 2001; 82: 311-5.

[17] Horiguchi J, Inami Y, Shoda T. Effects of long-term amantadine treatment on clinical symptoms and EEG of a patient in a vegetative state. *Clin. Neuropharmacol.* 1990; 13: 84-8.

[18] McMahon MA, Vargus-Adams JN, Michaud LJ, et al. Effects of amantadine in children with impaired consciousness caused by acquired brain injury: a pilot study. *Am. J. Phys. Med. Rehabil.* 2009; 88: 525-32.

[19] Giacino JT, Whyte J, Bagiella E, Kalmar K, Childs N, Khademi A, Eifert B, Long D, Katz DI, Cho S, Yablon SA, Luther M, Hammond FM, Nordenbo A, Novak P, Mercer W, Maurer-Karattup P, Sherer M. Placebo-controlled trial of amantadine for severe traumatic brain injury. *N. Engl. J. Med.* 2012 Mar 1; 366(9):819-26.

[20] Liepert J. Pharmacotherapy in restorative neurology. *Curr. Opin. Neurol.* 2008; 21: 639-43.

[21] Clauss RP, Gu¨ ldenpfennig WM, Nel HW, et al. Extraordinary arousal from semi-comatose state on zolpidem: a case report. *S Afr. Med. J.* 2000; 90: 68-72.

[22] Clauss R, NelW. Drug induced arousal from the permanent vegetative state. *NeuroRehabilitation* 2006; 21: 23-8.

[23] Shames JL, Ring H. Transient reversal of anoxic brain injury related minimally conscious state after Zolpidem administration: a case report. *Arch Phys Med Rehabil* 2008; 89: 386-8.

[24] Cohen L, Chaaban B, Habert MO. Transient improvement of aphasia with zolpidem. *N Engl J Med* 2004; 350: 949-50.

[25] Singh R, McDonald C, Dawson K, et al. Zolpidem in a minimally conscious state. Brain Inj 2008; 22: 103-6.

[26] Whyte J, Myers R. Incidence of clinically significant responses to zolpidem among patients with disorders of consciousness: a preliminary placebo controlled trial. *Am J Phys Med Rehabil* 2009; 88: 410-8.

[27] Sara` M, Sacco S, Cipolla F, et al. An unexpected recovery from permanent vegetative state. *Brain Inj.* 2007; 21: 101-3.

[28] Sara` M, Pistoia F, Mura E, et al. Intrathecal baclofen in patients with persistent vegetative state: two hypotheses. *Arch. Phys. Med. Rehabil.* 2009; 90: 1245-9.

[29] Taira T, Hori T. Intrathecal baclofen in the treatment of post-stroke central pain, dystonia, and persistent vegetative state. *Acta. Neurochir. Suppl.* 2007; 97: 227-9.

[30] Taira T. Intrathecal administration of GABA agonists in the vegetative state. *Prog. Brain Re.s* 2009; 177: 317-28.

[31] Georgiopoulos M, Katsakiori P, Kefalopoulou Z, Ellul J, Chroni E, Constantoyannis C. Vegetative state and minimally conscious state: a review of the therapeutic interventions. Stereotact *Funct. Neurosurg.* 2010; 88 (4):199-207.

[32] Cooper EB, Scherder EJ, Cooper JB: Electrical treatment of reduced consciousness: experience with coma and Alzheimer's disease. *Neuropsychol. Rehabil.* 2005; 15: 389–405.

[33] Schiff ND, Giacino JT, Kalmar K, Victor JD, Baker K, Gerber M, Fritz B, Eisenberg B, Biondi T, O'Connor J, Kobylarz EJ, Farris S, Machado A, McCagg C, Plum F, Fins JJ, Rezai AR: Behavioural improvements with thalamic stimulation after severe traumatic brain injury. *Nature* 2007; 448: 600–603.

[34] Canavero S, Massa-Micon B, Cauda F, Montanaro E: Bifocal extradural cortical stimulation-induced recovery of consciousness in the permanent post-traumatic vegetative state. *J. Neurol.* 2009; 256: 834–836.

[35] Kanno T, Morita I, Yamaguchi S, Yokoyama T, Kamei Y, Anil SM, Karagiozov KL: Dorsal column stimulation in persistent vegetative state. *Neuromodulation* 2009; 12: 33–38.

In: Chronic Disorders of Consciousness ISBN: 978-1-62618-940-9
Editors: R. S. Calabrò and P. Bramanti

Chapter 7

Rehabilitation and Nursing, Caregivers and Psychological Support in Chronic Disorders of Consciousness

Silvia Marino, Annalisa Baglieri, Francesco Corallo, Rosaria De Luca, Letteria Spadaro, Rocco Salvatore Calabrò and Placido Bramanti
IRCCS Centro Neurolesi "Bonino-Pulejo"

Abstract

The major disorders of consciousness include Coma, Vegetative State, and Minimally Conscious State. The care and the management of patients with Disorders of Consciousness has raised several important ethical, social and medical issues. It is evident that a uniform ethical framework needs to guide clinicians and caregivers in terms of clinical outcome, prognosis, and medical management. Over the last 10 years, specific recommendations for the assessment and rehabilitation of patients with disorders of consciousness have been promulgated by neurology and neurorehabilitation organizations to provide some guidelines for care.

In the present chapter, we describe the good clinical practice used in nursing and rehabilitation to provide the optimal recovery of these patients, through a multidisciplinary approach. Subsequently, we focus on the importance of the psychological support, in order to improve the quality of life of the patients' caregivers. The family is a critical and fundamental aspect in the management of these patients and it should be considered as integral part of care in the future guidelines.

Introduction

The major Disorders Of Consciousness (DOC) include Coma, Vegetative State (VS), and Minimally Conscious State (MCS). Coma is a state of complete unconsciousness from which the patient cannot be aroused. The eyes remain closed, there are no sleep-wake cycles on Electroencephalography (EEG), and there is no evidence of purposeful behavior or intelligible communication. When coma is caused by acquired brain injury (i.e., trauma, hypoxic-ischemic events, metabolic disorders), it usually evolves into VS or a conscious state within 2 to 4 weeks. In VS, the patient shows spontaneous eye opening, evidences of sleep-wake cycles, but no sign of environmental or self-awareness, and no ability to interact with others. The VS may be unchanged and so it may represent a permanent state of unconsciousness, or it may evolve in higher levels of consciousness, such as MCS. This condition has been defined as "a condition of severely altered consciousness in which minimal but definite behavioral evidence of self or environmental awareness is demonstrated" [1]. Evidence of awareness may be a sustained visual pursuit and gestural yes/no responses to simple commands, or movements and affective behaviors occurring in relation to a specific environmental events (i.e. smiling for a family photo).

Nursing Care and Rehabilitation in Patients with Disorders of Consciousness

The Royal College of Physicians Guidance on VS describes several fundamental steps that should be taken in the care of patients: adequate nutrition, usually via a percutaneous endoscopic gastrostomy tube; good skin care; joint movements to minimize contractures; management of spasticity (see Appendix), suction to avoid aspirations; management of bladder and bowel incontinence; care for dental and oral hygiene. It is also important to keep the family fully informed of the prognosis and any developments. Any decisions regarding treatment or long term rehabilitation are established by a physician's clinical judgment. The American Academy of Neurology (AAN) has reported certain fundamental aspects of the care and management of the persistent VS (PVS) patient [2]. According to AAN, the artificial provision of nutrition and hydration is considered one of the primary aspects of the medical treatment and may be discontinued in accordance with the principles and practices governing the withholding and withdrawal of other forms of medical treatment; it is good medical practice to initiate the artificial provision of fluids and nutrition when the patient's prognosis is uncertain, and to allow for the termination of treatment at a later date, when the patient's condition becomes hopeless. Development of interventions for treatment and rehabilitation of patients with DOC is a crucial challenge for current and future generations of neuroscientists. The management of PVS patients is an extremely difficult task for relatives and society in general, and these cases are usually considered hopeless. Although current treatments promoting recovery in such cases are extraordinarily limited, the findings of the one Cuban Group for Study of DOC, suggest new medical, ethical and practical implications for the diagnosis and management of PVS patients [3]. Indeed, the aforementioned Cuban Group is developing several research protocols to search for preserved residual brain and autonomic functions in PVS and MCS cases [4-11]. Recently, consensus-based recommendations for management of patients in the VS and MCS have been developed and disseminated in neurology and neurorehabilitation.

The care of patients with DOC has raised several important ethical, social and medical issues [12-15]. The allocation of healthcare resources is an important issue because DOC patients are maintained alive artificially and require prolonged care [16]. It is evident that a uniform ethical framework needs to guide clinicians and caregivers in terms of clinical outcome, prognosis, and medical management. The functions of the clinical management plan should aim to maintain and to improve the clinical state of the patient. In this context, it is fundamental the presence of a multidisciplinary clinical team, medical and paramedical, for preventing secondary complications and to provide the clinical and physical environment for the optimal recovery. The main complications requiring careful monitoring and appropriate preventive and treatment programmes are: increased muscle tone leading to contractures and permanent deformities, bladder and renal tract complications (infections, bladder stone and contracted bladder), bowel complications (constipation or diarrhea), under nutrition, respiratory infections, stress ulceration, deep vein thrombophlebitis, decubitus ulceration, heterotopic ossification, family dynamics and complications of medications. The following preventive options are available to the clinical team: negate and/or remove nociceptive environmental stimulating factors; correct positioning of the patient in bed using appropriate aids and techniques; provision of an appropriate supportive seating systems; standard chairs rarely provide sufficient support; emphasis should be placed on providing good support at the trunk, neck and head with specific emphasis being placed on keeping the hips, knees and ankles maintained at right angles; passive exercises and movement exercise on a regular basis; standing passive devises, prosthetic management.

Therefore, the principles of good nursing care are essential and include: optimal nutrition and hydration, maintenance of skin turgor; treatment of anaemia; maintenance of good oral hygiene, prevention of aspiration of food or fluids, prevention and reduction of the gastro-esophageal reflux, prevention of the higher risk of inhalation pneumonia associated with naso-gastric tube feeding, tracheostomy decannulation management, posture and position control, positioning addressed to maximize pulmonary drainage, appropriate transfer techniques to avoid shearing forces; appropriate support systems, e.g., bed, mattress, seating, wheelchair.

In the rehabilitation setting, the primary goals of the assessment in patients with DOC are to establish an accurate diagnosis, to develop a valid prognosis, and to design an appropriate treatment plan. The rehabilitation of patients with DOC is laced with challenge and uncertainty. Over the last 10 years, specific recommendations for the assessment and rehabilitation of patients with DOC have been promulgated by neurology and neurorehabilitation organizations to provide some guidelines for care [16, 17]. Standards of care have not been established to guide rehabilitative treatment in patients with DOC. Pharmacologic interventions, sensory stimulation, physical management procedures, hyperbaric oxygen therapy (HBO), and deep brain stimulation (DBS) have all been utilized for rehabilitative purposes. No treatment has been empirically shown to be clinically effective in restoring consciousness, cognition, or functional capacity. There is, however, supportive evidence that some interventions may promote recovery, particularly in patients with MCS.

Over the last few years, the Sensory Stimulation (SS) has played an increasingly important role in early rehabilitation therapy. The stimulation therapy consists of auditory, tactile, olfactory, gustatory and Kinesthetic procedures. SS is the most widely studied rehabilitative treatment for patients with DOC, although most published reports are based on uncontrolled case series [18]. Published data suggest that unimodal SS, multimodal SS, and sensory regulation are unproven treatments for reducing duration of coma or improving functional outcome in patients in coma, VS, and MCS [19].

Among the most promising instruments for the SS, the "Neurowave" (Khymeia group of Padua, Italy) is a technologically advanced clinical tool that allows the automatic administration of sensory stimuli (images, movies, sounds and voices, including specially patient-specific materials) and the simultaneous recording and monitoring of the physiological signals (figure 1).

Physical rehabilitation procedures, including passive range of movement, positioning protocols, and hygiene management, are routinely utilized in inpatient brain injury programs and are often required by accrediting agencies. Nevertheless, empirical evidence supporting their effectiveness is sparse and they remain unproven for improving physical, cognitive, or functional outcome in patients functioning at Rancho levels II and III [19].

HBO has been heavily touted by some of its proponents and appears to be increasingly utilized by consumers.

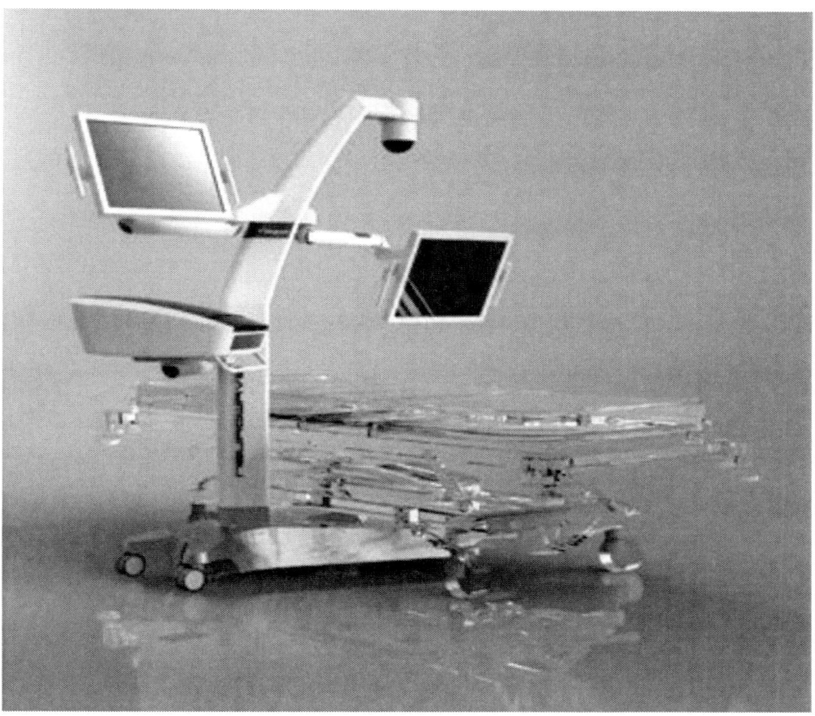

Figure 1. Shows the innovative Neurowave platform.

One well-controlled randomized trial suggested that HBO is effective in reducing mortality in patients with Glasgow Coma Scale (GCS) scores of 9 or greater when it is administered within 24 hours of injury [20]. The same study suggested that HBO is ineffective in reducing severity of disability (i.e., the number of cases with severe disability or VS at 12 months post-injury).

DBS, a well-recognized treatment for control of tremor in Parkinson's disease, has been used to facilitate recovery of consciousness in patients diagnosed with VS [21, 22]. In this procedure, electrodes placed in brain stem or diencephalic structures deliver pulses of electrical activity in an effort to stimulate the reticular system. The resultant heightening of reticular function is presumed to increase cortical activation and, ultimately, promote behavioral responsiveness. Despite published claims that DBS can facilitate recovery of consciousness in patients in VS, it is currently considered an unproven intervention because of significant flaws in the design of the supportive studies. Moreover, severely brain-damaged patients can get relevant advantages, by using the Augmentative and Alternative Communication (AAC). It is an innovative and promising rehabilitative tool based on the patient selection and choice of images or symbols showed on the computer screen, through the eye pursuit or the other residual sensory-motor abilities. To date, the AAC is successfully applied in Locked-in patients, who may present less unstable responses [23].

Therefore, rehabilitative management of patients with DOC continues to be plagued by many unanswerable questions. In the absence of clear-cut guidelines, it is critical that clinicians adopt a systematic approach from the assessment to the rehabilitative treatment.

In this field, it should be interesting to consider and to develop these stimulation interventions as integral part of nursing into a multidisciplinary management of patients with DOC.

Caregivers and Psychological Support

In the caring of patients with DOC a particular attention has to be directed to the family and especially to the informal caregiver. The relatives are the first resource of assistance for the patient. If the need is long-lasting, as in chronic diseases, the duties of care can become a permanent engagement over time. If all the family is involved in the DOC care, the caregiver attends to the DOC person with particular attention. The informal caregiver feels the responsibility of caring in the relationship with DOC patients more than other family members. The formal caregiver gets benefits for the patient assistance.

In this section, we are treating the psychological aspects of the family and the informal caregivers, in particular, after the diagnosis of DOC for their family member.

The management of patients with DOC usually presents medical, ethical and psychological challenges for patients, clinicians and caregivers too. Health professionals need to be responsive to the reactions of the families. The unpredictability and severity of brain injury of DOC may usually cause deep emotional distress for the family. The stages of the disease course (critical / acute / post-acute / chronic) correspond to as many important moments for relatives. In the first two phases, the fear for the life of their suffering family member is very strong. In the other two phases, the family may have some difficulties to understand a patient who is severely neurologically impaired. A DOC person is not able to assume the social role as child, husband/wife, parent, brother/sister and to express clearly emotions and thoughts. Mobility, communication and cognitive processes disappear, so that the DOC person become completely dependent.

The psychological reactions of relatives, when a family member is a person with an altered conscious level, have been investigated in few clinical researches. All the authors underline the stress of the family after diagnosis [24, 25]. Before the relatives are able to accept the situation, they may usually experience a range of reactions including shock, anxiety, guilt, denial, depression and hostility towards staff caring for the patient [26]. In most cases the elaboration of disease unfolds in physiological times and modes, enabling the caregiver to react to the separation and to resume their own daily life. In some cases, the caregivers are unable to react and remain imprisoned in prolonged grief with the difficulty of processing the past and the future in the new condition of life [26]. This pain may affect psychological, physical and relational well-being of the caregiver. If the hope of a recovery fades, they may also experience psychosomatic disturbances, insomnia and the loss of appetite, anxiety, anger and aggressiveness during the months following the traumatic patient's event together with grief [27]. Even if these patients are hospitalized, most of their family members take care of them every day for several hours a day [28]. The relatives of the same patient may also experience different psychological reactions. For example, the patient's partner may be ready to begin accepting the situation earlier than the patient's parents. They could perceive a reduction in visiting by the partner as a betrayal of the patient, and they may even believe it to be detrimental to the patient's chances of recovery. Such conflicts can only add stress to the distress experienced by the family.

In particular, the relatives of VS patients live a paradoxical reality: their family member is both present and absent; alive and dead [24, 25]. The patients are not dead, they have some, involuntary or few voluntary, movements and are more independent of vital support then in intensive therapy [25]. For this reason, the caregivers often deny the happening or develop fantasies concerning the reality of the incident and its possible resolution (supported by spastic or reflexes movements interpreted as signs of improvement) [29]. The relatives could report that the patient had important responses when these have not been observed by the professionals who have been assessing the patient and, commonly fail to understand the nature of the observed behaviour. Sometimes patients' behaviours are really different when their relatives are present. For this reason, the staff should observe the patient with the relatives and involve them in the patient evaluation. This is important for the caregivers to better understand the patient's behaviours and the level of awareness [30] and to strengthen the relationship with the staff. In all the phases of DOC, the families and healthcare workers may have a complex relationship. In fact, the uncertainty and unreliability of diagnosis and prognosis is likely to be a strong feature in discussions between the health professionals and the family. Some relatives may be denying of the extent of the patient's impairments and the poor prognosis [27, 30]. A clear communication of short-term goals of treatment and the involving of the caregivers in the cares are important to help families to understand and to participate in the course of treatment, to cope with this particular pathological clinical condition [31].

Many patients do not emerge from VS, and recovery is rare more than 12 months after TBI and more than 6 months after non-traumatic brain injury [32, 33]. The hope is an important dimension in the care of DOC caregivers.

As the DOC becomes a chronic condition, the care that families provide needs to be in the context of a specialized team approach, and families require specialized care and support at home or in an institution. In Italy, United Kingdom and, France there are specialist care facilities for people with severe acquired brain injury, targeting services to the actual needs of this client group. However, the care at home is delivered by family who needs specialized support to be able to manage the care of their family member at home and to maintain the intensity of care required. If the family is adequately supported, they can make a very positive contribution to the care of people in DOC, as they are the people with the knowledge of the person's personality, values and preferences before the injury [34]. Assessing the needs represents a first step to gain insights into the family members' daily condition and to plan targeted interventions.

Studies on informal caregivers of patients with chronic pathologies showed the impact of care-giving in terms of burden and distress [35]. The caregiver may feel the changes related to the disease as a physical and psychosocial burden with direct and indirect financial expenses. The tasks related to the care can be exhausting but especially psychological well-being is affected. Garlo and collaborators observed that burden was associated with perceived needs of support during daily tasks and not with the patient's objective needs for assistance; the burden may be influenced by the adaptation to care-giving role [36]. For example, the caregivers of long-term stroke survivors reported adverse and long-standing difficulties in emotional health, social activities and family relationships [37]. Becoming a caregiver is a personal path. To date, few studies have investigated the experience of the family members who take care of patients with DOC. In particular, few researches have studied how families constantly cope these pathologies, what coping strategies are developed in the everyday life, if the high levels of depression and anxiety modify over time, and if the place of care chosen for the patient, at home or in a long-term hospital structure, influences the emotional and psychological state of the family members.

The duration and the uncertainty of the disease may produce psychological and physical stress with an important economic burden [27]. Some time the caregiver lose their jobs for the need to move to cities where there is a specialized unit for the patients or a partner may suddenly become the sole earner in the family at a time when they have also lost their main source of emotional support. Many caregivers refers also the loss of interests and hobbies, limited social relationship associated with a feeling of loneliness [33]. The grief may become a chronic condition for the caregiver [38, 39]. Nevertheless, the changes in roles may actually have a positive psychological outcome, at least in the short term: some relatives appear to draw strength from developing a caring role for the patient, and may form a new social network with relatives of other patients in a similar state.

Chiambretto and collaborators have presented the results of a survey conducted on 30 family members of patients in VS, 25 of whom were long-term hospital inpatients and 5 home-based. Results are compared with the findings which emerged from a longitudinal study carried out on 16 family members tested in the year 2000 and re-evaluated, utilizing the same assessment tools, 5 years later. The results reveal a picture of significant emotional and psychophysical distress, high levels of anxiety and depression and the use mainly of situation-focused coping strategies. Family relations appear little satisfying even if over time they tend to improve. The comparison at 5 years reveals levels of anxiety, depression and emotional burnout superior to the norm and almost unchanged despite the course of the time. The comparison is slightly better for those who take part in mutual help groups under the guidance of a psychologist; these subjects tend to use coping strategies more centred on the expression of their emotions. Group therapy, when its goal is to help caregivers share their own personal experience concerning the caring relationship, appears to facilitate the development of new strategies and resources to cope with daily living [26].

In the international literature, there are a lot of studies about caregiving-related problems, but the assessment tools are generally disease-specific, often only about the psychological burden, frequently characterized by unsatisfactory psychometric criteria and, above all, sometimes require the use of an excessive number of questions [40]. On the basis of these considerations, the aim of Rossi and collaborators was to develop a reference model for needs analysis based on the findings of international research into caregiving-related problems and the results obtained in Italy: Economic, Social, Psychological and Educational Requirements [41]. The model includes a specific questionnaire, the Family Strain Questionnaire (FSQ) that makes possible to detect the psychological and social problems of caregivers looking after patients with different diseases. This instrument is both useful and economic, and the authors believe that it may allow the most rational choice of socio-psychological interventions and the use of local resources [33]. The FSQ consists of a form for recording demographic data; a semi-structured interview designed to obtain a description of the disease and the way in which it is managed, and to identify any economic problems and the use of health and social welfare resources; 35 dichotomous items (yes/no) covering five factors which explain 41.5% of the total variance: (1) emotional burden; (2) problems of social involvement; (3) the need for knowledge of the disease; (4) the quality of family relationships; and (5) thoughts about death; and nine non-factorial dichotomous items for clinical use concerning the caregivers attribution of problems relating to social stigma and affective exchange to her- or himself and her or his patient. The FSQ can be administered relatively quickly (on average, 20 minutes) and is characterized by good psychometric properties.

On the basis on above-described considerations, the family is a critical and fundamental aspect, that should be included in the global care of patients with

DOC, and as integral part of the future guidelines on management of these patients.

Appendix

Management of Spasticity: An Overview

Clinically spasticity is defined as velocity dependent resistance to stretch, where a lack of inhibition results in excessive contraction of the muscles, ultimately leading to hyperflexia (overly flexed joints). Spasticity is a common and disabling symptom for many patients with upper motor neuron dysfunction, resulting from interruption of inhibitory descending spinal motor pathways, and although the pathophysiology of spasticity is poorly understood, the final common pathway is overactivity of the alpha motor neuron. Spasticity is assessed by feeling the resistance of the muscle to passive lengthening in its most relaxed state. A spastic muscle will have immediately noticeable, often quite forceful, increased resistance to passive stretch when moved with speed and/or while attempting to be stretched out, as compared to the non-spastic muscles in the same person's body (if any exist). The Asworth scale is the most common used tool to assess spasticity.

Therapy for spasticity is symptomatic with the aim of increasing functional capacity and relieving discomfort. Any approach to treatment should be multidisciplinary, including physical therapy, and possibly surgery, as well as pharmacotherapy. Physical therapy is a mainstay of treatment for spasticity and it refers to a range of physical treatments, designed to reduce muscle tone, maintain or improve range of motion and mobility, increase strength, and improve care and comfort. The choice of treatments is individualized to meet the needs of the person (and/or - as in DOC- the caregivers' need) with spasticity. The use of oral medication to treat spasticity is indicated when stiffness, spasms, or clonus interfere with daily functioning or with sleep. Baclofen is a derivative of gamma-aminobutyric acid (GABA) and produces its effects by activating the $GABA_B$ receptor. Indeed, it acts on the central nervous system to relax muscles by decreasing stretch reflexes, the rate of muscle spasms and clonus, pain, and tightness and improves range of motion. It can either be taken by mouth as a tablet or delivered via a catheter into the space surrounding the spinal cord (the intrathecal space—please see the later section on ITB Therapy). Side effects may include sedation, drowsiness, weakness, decreased muscle tone, confusion, fatigue, nausea, dizziness, and more difficulty in controlling seizures in people with epilepsy. Baclofen may also increase the blood pressure-lowering effects of other medications. Suddenly stopping baclofen may cause seizures, hallucinations, and rebound spasticity. Benzodiazepines are a group of drugs that act on the central nervous system to relax muscles, temporarily decreasing spasticity. Diazepam is the most commonly used benzodiazepine in the treatment of spasticity. It enhances the effect of the neurotransmitter GABA by binding to the benzodiazepine site on the $GABA_A$ receptor leading to central nervous system depression. Because of their sedative effects, benzodiazepines are given most often at night, even in DOC - if possible- so to not alter circadian cycle. Benzodiazepines may be addictive, and, with long-term use, withdrawal symptoms may occur if the drug is stopped too quickly. These drugs should not be taken in combination with other central nervous system depressants. Clonidine (which is available as a tablet or a patch worn on the skin) and tizanidine are two imidazolines, and centrally acting α_2 adrenergic agonist; thus, they reduce spasticity through their action on the CNS. Interestingly, these drugs typically cause less muscle weakness than do oral baclofen and benzodiazepines. Gabapentin, and its new derivate Pregabalin, are GABA-derivate drugs typically used in the treatment of seizures and neuropathic pain; however, they are also used in the treatment of spasticity, especially caused caused by MS and spinal cord injury.

As previously mentioned, baclofen can be delivered directly into the thecal space with a three-part system made up of a programmable pump with a reservoir or storage area for the drug; a clear, flexible silicone tube or catheter; and a programming device. This delivery method is called intrathecal therapy (ITB). Because the drug is administered directly to its site of action- within the spinal cord -much less baclofen is needed than if the drug were taken by mouth and more of the drug actually reaches the nerve cells where it is needed. Because less of the medication is needed (often 100 times less), the side effects of ITB, including drowsiness and sedation, are much milder than when baclofen is taken by mouth. ITB is especially useful for treating spasticity in people with cerebral palsy and is most effective for spasticity of the lower limbs. Candidates for ITB have disabling spasticity that does not respond to conservative treatment with medications or have intolerable side effects at therapeutic doses. To determine whether ITB is expected to produce a helpful response, the patient undergoes an ITB Therapy screening test. The screening test for ITB requires the administration of a test dose of baclofen (typically with 50 mcg, usually not to exceed 100 mcg) via lumbar puncture into the thecal space. Peak effect of the drug usually occurs within four hours. Patients who respond positively to the test dose can be considered for chronic ITB treatment. During screening, patients must be monitored closely in a fully equipped and staffed setting, due to the rare risk of life-threatening side effects. Once it has been determined that ITB is likely to be effective in treating spasticity, a surgeon performs an operation to place a permanent battery-powered pump into place. The dose of baclofen can be adjusted whenever necessary by reprogramming the pump in the doctor's office. The pump also contains a programmable alarm that beeps softly when the reservoir is low or the batteries need replacing. The reservoir is refilled by injection as needed, usually every 1 to 3 months. When the batteries run low, about every 4 to 5 years, the entire pump is replaced. About 5% of people with ITB may develop infections that require temporary removal of the pump. Other equipment-related risks include pump failure, tube kinking or breakage, or movement of the catheter (dislodgement) so that the baclofen no longer reaches the intrathecal space. Mechanical defects or failure to refill the pump reservoir can lead to sudden interruption of baclofen treatment. This can cause a life-threatening withdrawal syndrome. Although it rarely happens, it is possible for the person receiving too much ITB medication (overdose), which, if misdiagnosed, may lead to coma and death.

Botulinum toxin is a neurotoxin that is produced by the bacterium responsible for botulism, *Clostridium botulinum*, working by weakening the muscles and stopping them from contracting very well, which can lead to paralysis. BTX-A is nowadays commonly used to treat spasticity. However, often, very large doses of the injection are given into the 'spastic' muscles to obtain amelioration comfort, mobility or hygiene.

Rhizotomy is a term chiefly referring to a neurosurgical procedure that selectively destroys problematic nerve roots in the spinal cord, most often to relieve the symptoms of neuromuscular conditions such as spastic diplegia and other forms of spastic cerebral palsy. The selective dorsal rhizotomy (SDR) for spastic cerebral palsy has been the main use of rhizotomy for neurosurgeons specialising in spastic CP since the 1980s; in this surgery, the spasticity-causing nerves are isolated and then targeted and destroyed. The sensory nerve roots, where spasticity is located, are first separated from the motor ones, and the nerve fibres to be cut are then identified via electromyographic stimulation; The ones producing spasticity are then selectively lesioned with tiny electrical pulses.

In conclusion, it is important to underline that the complex and multidisciplinary spasticity treatment has specif goals that should overcome the inappropriate caregivers expectations about the effectiveness of treatment.

Therefore, ongoing communication and agreement between caregivers and healthcare professionals regarding the goals of treatment are extremely important. Thus the goals that are commonly developed in the treatment of spasticity include:

1. Relieve the signs and symptoms of spasticity
2. Reduce pain, frequency of spasms, or irritating stimuli
3. Improve gait, hygiene, activities of daily living, or ease of care
4. Reduce problems with passive function, that is, the functions provided by the caregiver, such as dressing, feeding, transfer, and bathing
5. Improve voluntary active motor function, that is, the behaviors and functions that are under the patient's control, such as reaching for, grasping, moving, and releasing an object.

References

[1] Giacino, JT; Trott, CT. Rehabilitative Management of Patients With Disorders of Consciousness. Grand Rounds. *Journal of Head Trauma Rehabilitation* 2004; 19(3): 254–265.

[2] American Academy of Neurology. Position of the American academy of neurology on certain aspects of the care and management of the persistent vegetative state patient. *Neurology* 1989; 39: 125-126.

[3] Machado, C; Estévez, M; Rodríguez, R; Pérez-Nellar, J; Gutiérrez, J; Carballo, M; et al. A Cuban Perspective on Management of Persistent Vegetative State. *MEDICC Review* 2012; 14(1): 44-8.

[4] Machado, C; Estévez, M; Rodríguez, R; Carballo, M; Pérez–Nellar, J; Gutiérrez, J; et al. Are Persistent Vegetative State Patients Isolated from the Outer World? *Journal of Functional Neurology, Rehabilitation and Ergonomics* 2011; 1(2): 357–78.

[5] Machado, C; Estévez, M; Pérez–Nellar, J; Gutiérrez, J; Rodríguez, R; Carballo, M et al. Autonomic, EEG, and Behavioral Arousal Signs in a PVS Case After Zolpidem Intake. *Canadian Journal of Neurological Sciences* 2011; 38(2): 341–4.

[6] Machado, C; Estévez, M; Rodríguez, R; Pérez–Nellar, J; Silva, S; Loubinoux, I; et al. Wakefulness and loss of awareness: Brain and brainstem interaction in the vegetative state. *Neurology* 2010; 75(8): 751.

[7] Machado, C; Pérez–Nellar, J; Rodríguez, R; Scherle, C; Korein, J. Emergence from minimally conscious state: insights from evaluation of Post-traumatic confusion. *Neurology* 2010; 74(14): 1156–7.
[8] Machado, C; Leisman, G. Towards an effective. Definition of death and disorders of consciousness. *Review in the Neurosciences* 2009; 20(3– 4): 147–50.
[9] Machado, C; Rodríguez, R; Carballo, M; Korein, J; Sánchez-Catasus, C; Pérez, J; et al. Brain anatomy, cerebral blood fl ow, and connectivity in the transition from PVS to MCS. *Review in the Neurosciences* 2009; 20(3–4): 177–80.
[10] Machado, C; Korein, J. Persistent vegetative and minimally conscious states. *Review in the Neurosciences* 2009; 20(3–4): 203–20.
[11] Machado, C; Korein, J; Aubert, E; Bosch, J; Alvarez, MA; Rodríguez, R; et al. Recognizing a mother's voice in the persistent vegetative state. *Clinical EEG and Neuroscience* 2007; 38(3): 124–6.
[12] Racine, E; Rodrigue, C; Bernat, JL; Riopelle, R; Shemie, SD. Observations on the ethical and social aspects of disorders of consciousness. *Canadian Journal of Neurological Sciences* 2010; 37(6): 758-768.
[13] Bacon, D; Williams, MA; Gordon, J. Position statement on laws and regulations concerning life-sustaining treatment, including artificial nutrition and hydration, for patients lacking decision-making capacity. *Neurology* 2007; 68(14): 1097-1100.
[14] Bernat, JL. Ethical issues in the management of patients with impaired consciousness. *Handbook of Clinical Neurology* 2008; 90: 369-382.
[15] Bernat, JL. Chronic consciousness disorders. *Annual Review of Medicine* 2009; 60: 381-392.
[16] The Multi-Society Task Force Report on PVS. Medical aspects of the persistent vegetative state. *The New England Journal of Medicine* 1994; 330:1499–1508, 1572–1579.
[17] Aspen Neurobehavioral Conference Workgroup. The Aspen Neurobehavioral Conference Consensus Statement on the Vegetative and Minimally Conscious States. In: *Proceedings of the Aspen Neurobehavioral Conference*. Given Biomedical Institute; Aspen, Colo; 1996; 19–23.
[18] Giacino, JT. Sensory stimulation: theoretical perspectives and the evidence for effectiveness. *Neurorehabilitation* 1996; 6: 69–78.
[19] Giacino, JT. Rehabilitation of patients with disorders of consciousness: An evidence-based review. In: High W, Hart K, eds. *Rehabilitation*

Interventions Following TBI: State of the Science. New York: Oxford University Press. In press.

[20] Rockswold, GL; Ford, SE; Anderson, DC; Bergman, TA; Sherman, RE. Results of a prospective randomized trial for treatment of severely brain-injured patients with hyperbaric oxygen. *Journal of Neurosurgery* 1992; 76: 929–934.

[21] Kanno, T; Kamei, Y; Yokoyama, T; Jain, VK. Neurostimulation for patients in vegetative status. *PACE*. 1987; 10: 207–208.

[22] Tsubokawa, T; Yamamoto, T; Katayama, Y; Hirayama, T; Maejima, S; Moriya, T. Deep brain stimulation in a persistent vegetative state: follow up results and selection of candidates. *Brain Injury* 1990; 4(4): 315–327.

[23] Hill, K. Advances in augmentative and alternative communication as quality-of-life technology. *Physical Medicine and Rehabilitation Clinics of North America* 2010; 21(1), 43-58.

[24] Lezak, MD. Brain damage is a family affair. *Journal of Clinical and Experimental Neuropsychology* 1988; 10: 111-123.

[25] Stern, J; Sazbon, L; Becker, E; Costeff, H. Severe behavioural disturbance in families of patients with prolonged coma. *Brain Injury* 1988; 21: 256-62.

[26] Chiambretto, P; Vanoli, D. Family reactions to the vegetative state: a follow-up after 5 years. *Giornale Italiano di Medicina del Lavoro ed Ergonomia* 2006; 28: 15–21.

[27] Tzidkiahu, T; Sazbon, L; Solzi, P. Characteristic reactions of relatives of post-coma unawareness patients in the process of adjusting to loss. *Brain Injury* 1994; 8: 159-165.

[28] Tresch, D; Sims, F; Duthie, E; Goldstein, M. Patients in Persistent Vegetative State attitudes and reactions of family members. *Journal of the American Geriatrics Society* 1991; 39: 17-21.

[29] Bowlby, J. *Attaccamento e perdita: la perdita della madre.* Torino: Bollati Boringhieri; 1983.

[30] Jacobs, HE; Muir, CA; Cline, JD. Family reactions to persistent vegetative state. *The Journal of Head Trauma Rehabilitation* 1986; 1(1): 55-62.

[31] Crawford, S. Psychological of patients, families and health professionals [online]. 2004 [cited 2012 August 7] Available from: http//www.timeoutintensiva.it/atchivio/Crawford.pdf

[32] Royal College of Physicians. *The Vegetative State: Guidance on Diagnosis and Management.* London: Royal College of Physicians; 2003.

[33] Chiambretto, P; Rossi Ferrario, S; Zotti, AM. Patients in a persistent vegetative state: caregiver attitudes and reactions. *Acta Neuroligica Scandinavica* 2001; 104: 364-68.
[34] Colin-Thomé, D; Craig, G. CCG commissioning. Look after number one. *Health Service Journal* 2012; 122(6298): 26-7.
[35] Ho, SC; Chan, A; Woo, J; Chong, P; Sham, A. Impact of caregiving on health and quality of life: a comparative population-based study of caregivers for elderly persons and noncaregivers. *Journals of Gerontology Series A: Biological Sciences and Medical Sciences* 2009; 64(8): 873-9.
[36] Garlo, K; O'leary, JR; Van Ness, PH; Fried, TR. Burden in caregivers of older adults with advanced illness. *Journal of the American Geriatrics Society* 2010; 58: 2315–22.
[37] Anderson, CS; Linto, J; Stewart-Wynne, EG. A population based assessment of the impact and burden of caregiving for long-term stroke survivors. *Stroke* 1995; 26: 843–9.
[38] Guarnerio, C; Prunas, A; Della Fontana, I; Chiambretto, P. Prevalence and comorbidity of prolonged grief disorder in a sample of caregivers of patients in a vegetative state. *Psychiatric Quartely* 2012; 83(1): 65-73.
[39] Chiambretto, P; Moroni, L; Guarnerio, C; Bertolotti, G; Prigerson, HG. Prolonged grief and depression in caregivers of patients in vegetative state. *Brain Injury* 2010; 24(4): 581-8.
[40] Rossi Ferrario, S; Zotti, AM; Baroni, A; Cavagnino, A; Fornara, R. Emotional reactions and practical problems of the caregivers of hemodialysed patients. *Journal of Nephrology* 2002a; 15(1): 54-60.
[41] Rossi Ferrario, S; Baiardi, P; Zotti, AM. Assessment delle problematiche caregiving-correlate: il Family Strain Questionnaire. *Giornale Italiano di Medicina del Lavoro ed Ergonomia* 2001b; 23(1): 25-29.

In: Chronic Disorders of Consciousness ISBN: 978-1-62618-940-9
Editors: R. S. Calabrò and P. Bramanti

Chapter 8

Nutrition and Other Medical Problems in VS and MCS

***Daniela Floridia*[1], *Valeria Conti-Nibali*[1], *Rocco Salvatore Calabrò*[1], *Placido Bramanti*[1] *and Carlo Salviera*[2]**
[1] IRCCS Centro Neurolesi "Bonino-Pulejo", Messina, Italy
[2]Otolaringology Unit, Policlinico Universitario, Messina, Italy

Abstract

Enteral nutrition is the delivery of nutrients in liquid form directly into the stomach, duodenum, or jejunum and it is an integral part of the therapeutic approach in patients affected by vegetative and minimally conscious state. Indeed, artificial nutrition is of primary importance in ensuring adequate nutrients for the prevention of protein and caloric deficit responsible for the highest increase in therapeutic failure, medical complications and mortality. *Neurogenic dysphagia* or *preesophageal dysphagia* is difficulty in swallowing food caused by disease or impairment of the nervous system.Perhaps, the most clinically apparent complication of neurogenic dysphagia is pulmonary aspiration, which may manifest itself acutely as choking or coughing, respiratory distress, wheezing, gasping or gurgling, and tachycardia. However, chronic symptoms (particularly in silent aspiration), including weight loss, production of excessive oral secretions and aspiration pneumonia may be also present.

This chapter aims at evaluating the proper management of nutrition and other medical problems, with particular regards to dysphagia, in patients affected by chronic disorders od consciousness.

Artificial Nutrition

Introdution

Artificial nutrition is the key treatment in Vegetative State (VS) and Minimally Conscious State (MCS) patients even after the acute phase. These patients, which come from Intensive Care Units (Resuscitation and Neurosurgery) since are affected bysevere head trauma or vascular accidents, have complex outcomes.

Indeed, they are still tracheotomized, subjected to enteral and parenteral nutrition, have pathological postures due to decortication/decerebration, have severe spasticity and vegetative dysautonomy with profuse sweating, tachycardia and tachycardia.

Moreover, they are at high risk for developing diseases due to bed riddance that can, therefore, undermine the rehabilitative potential and increase overall hospitalization.

Furthermore, these subjects often suffer of metabolic imbalance, but artificially supplied nutrients can reverse catabolism, preserve and promote the integrity of the immune status thus reinforcing organism resistance.Adequate nutrition is, therefore, important to avoid complications, reduce hospitalization time, improve quality of life, and make the therapeutic path simpler and more effective.

Enteral Nutrition (EN) is an integral part of the therapeutic approach and plays an important role in outcome of the patient. It consists in direct administration of nutritional solutions into the gastroenteric tract. To date, it is the choice treatment since it allows direct nutrition through the physiological gastrointestinal system, thus conserving the trophism of the intestinal mucosa. Moreover, from a metabolic point of view, it is better tolerated and has a lower cost than total parenteral nutrition (TPN).

Artificial nutrition is of primary importance in ensuring the adequate nutrition for the prevention of protein and caloric deficit (PCD) responsible for the highest increase in therapeutic failure, complications and mortality.

Clinical and Laboratory Evaluation

The first approach of the nutritionist, within 24-48 hours from admission in the VS or SMC ward, is the assessment of the nutritional status and subsequent monitoring of the patient. The evaluation should begin with the nutritional anamnesis in order to highlight the presence of other diseases and/or treatments that may result in increased needs for energy and nutrients or request changes in the nutritional composition of the diet. The oral intake of foods is not easily assessable in such patients, and it is often considered inappropriate and therefore at nutritional risk when the intake is generally lower than 50% of the overall needs for a period greater than 7-10 days. The initial evaluation includes a careful examination and detection of anthropometric parameters. Among them, weight, height, and Body Mass Index (BMI), an important predictor of mortality in patients hospitalized for serious conditions, are considered first-level indicators.The visceral proteins do not constitute a specific indicator of nutrition since they can depend on a number of factors, whereas the evaluation of the immunological status through total lymphocytic count may be of some help. Repeated body weight measurement should be carried out at least weekly for the entire duration of hospitalization. Unintentional weight loss (clinical sign of negative caloric protein balance) of 5% in 3 months or 10% compared to normal weight before illness in the last six months is considered a risk factor for malnutrition. Other important predictors, indicated as second level, are the triceps skin fold (measured by the calliper of Harped en that provides an estimate of the lipid reserves) and the circumference of the arm (used for the measurement of the muscle reserve). The Impedance,- in particular the new piezoelectric Soft Tissue Analyser (STA), if weight, height and other anthropometric parameters are unknown, as is often the case for the patient in VS and SMC-, is a non-invasive, rapid and objective analysis capable of directly identifying cell mass in Kg and percentage content of extracellular water. Adequate nutrient supply is indicated by an increase in weight, oedema disappearance, increased muscle strength and work capacity, improved appetite and a sense of well-being. The of vitamins and trace elements status, the protein balance and protein metabolic rate, the evaluation of delayed hypersensitivity reaction (skin test) can be considered other valuable markers.

Total parenteral nutrition is indicated in the nutritional treatment of these subjects when it is impossible to maintain adequate nutrition, fluids and electrolytic balance through oral or enteral routes for a period exceeding 14 days or in case of non-availability of a peripheral venous access. It is the administration of energy substrates through a central venous catheter (CVC). In case of long-term CVC (more than two weeks) tunnelling through the subcutaneous tissue can be taken into consideration. This technique offers the advantage of distancing the point of emergence from the insertion point thus creating a natural barrier against bacterial or fungal contamination.

Swallowing Disorders

Introduction

A peculiar symptom severely affecting the rehabilitation project of VC/MCS patients is dysphagia. Neurogenic dysphagia it is closely associated with the severity of the neurological injury requires a careful integration between clinical care and rehabilitation leading to the prevention of fatal respiratory infections.

Indeed, the assessment, diagnosis and treatment of swallowing deficits have to be carried out in order to reduce the risk of developing serious pulmonary complications (aspiration pneumonia) and nutritional (protein-energy malnutrition, dehydration). In the early stages, a physician skilled in swallowing should carry out an accurate assessment also using specific diagnostic instruments to quantify the degree of deficit. Moreover, the collaboration with the specialist in nutrition and dietician will help identify the most appropriate diet for the patient.

Defining the Dysphagia Problem

The level of responsiveness is one of the most important parameters to be considered in the management of individuals dysphagia.To this end, the classification of the patient in class IV, according to the "Rancho Los Amigos Scale of Cognitive Function", is the minimum requirement for programming oral food intake. In addition to the altered state of consciousness, there are numerous issues that characterize the patient and interweave with the swallowing disorder. The VS or MCS patients often present neurovegetative crisis and important medical problems that strongly influence the choice of dysphagia treatment thus significantly narrowing the possibility of applying compensation methods. Unconscious patients may present different causes for dysphagia due to functional abnormalities, central and peripheral organic neurological problems, cognitive and behavioural alterations. Dysphagia in the patient in a VSmay be related to:

- Disorganization of the motor patterns involved in swallowing, within a broader framework of sensory, motor and cognitive abnormalities.
- Partially reversible alterations, linked to the presence of the tracheostomy tube and the duration of assisted ventilation.
- Very low levels of consciousness.
- Central neurological brain stem lesions responsible for reduced changes in motility, tone and oro-pharyngo-laryngeal sensitivity, and absent/reduced swallowing reflex and cough.
- Facial or jaw fractures, tooth extraction, changes in the mandibular articulation as possible consequences of pro-tracheal intubation and prolonged maintenance of supine posture and half-open mouth during the acute phase.
- Cranial nerve injury.
- Presence of archaic reflexes.
- Important disorders of head and trunk control
- Poor oral hygiene, with the presence of mycosis and inflammatory conditions that may interfere with the swallowing mechanism.

In VS/MCS patients many of the protective conditions for safe swallowing, such as raclage, voluntary coughing, swallowing on demand so to maintain a safe posture and the pneumo-phonic coordination, are not present.These elements are essential for a safety swallowing. Interestingly, in some patients it is possible to implement a pure swallowing reflex, in which the food in the mouth activates the process, in the absence of awareness or attention. Nonetheless, the swallowing phases that generally are more damaged are the preparatory phase, i.e. oral and pharyngeal ones, while the esophageal phase is greatly spared. More specifically, neurogenic dysphagia often presents with incoordination, slow and imprecise swallowing, prolonged oral transit, depressed swallowing reflexes, and reduced pharyngeal peristalsis.

Assessment of Dysphagia

A specific dysphagia protocol should be applied as soon as the patient is hospitalized, and requires the use of methods chosen according to the degree of consciousness of the patient. In fact, given the low/no cooperation of the patient, clinical assessment tools are initially the observation and stimulation of the patient. The clinical evaluation of dysphagia is usually performed by skilled healthcare professionalsin the management of dysphagia, usually a physician and/or speech therapist with specialized training in swallowing. It usually includes a bedside evaluation, i.e. Bedside Examination (BSE), motor function and signs and symptoms of dysphagia, the possible presence of pneumonia, unjustified weight loss, pulmonary status, current methods of nutrition and nutritional status compared with the previous state, the behavioural characteristics, with particular attention to the level of alertness, the appropriate position of the patient, the oral motor structures, general sensitivity and functions. The initial evaluation of swallowing requires a careful examination of both motor and sensory skills of all the mouth parts. Clinical examination should address, in particular, the labial competence, tongue motility and soft palate, the extent of drooling and areal space, cough and gag reflex, breathing patterns. The clinical evaluation generally consists of 2 parts. The first mainly includes the observation without the use of gustatory stimulus:

1. Posture and mobility related issues, possibility of control of the head and trunk.
2. Type of current breathing, evaluation for possible chest trauma, presence and type of tracheostomy tube, quality and quantity of secretions.
3. Oral hygiene.
4. Observation and manipulation of the lips, cheeks, jaw, floor of the mouth, tongue, oral film and cavity so to assess morphology, symmetry, muscle tone, sensitivity, any retractions of the oro-motor district.
5. Observation of possible spontaneous oral motility, and the appearance of pathological oral automatisms and reflexes after stimulation of the lips, cheeks and oral cavity.
6. Presence of possible spontaneous swallowing and their characteristics (multiple attempts, laryngeal lifting, pushing of the tongue).
7. Quality of voice and cough if issued.

The second part consists of the administration of small amounts of water and of gelled water, which is indicated for the low risk of inhalation and subsequent complications. The addition of methylene blue may be useful in the evaluation of symptoms such as cough and voice changes, and the possible detection of changes in oxygen saturation in the blood using transcutaneous oximetry.

In this evaluation it is possible to observe the appearance of automatisms such as the mouth opening, closure of the lips around a teaspoon, triggering of deglutition, the appearance of reflexes and lingual movements. Nevertheless, this assessment cannot exclude the presence of silent inhalation. The test with methylene blue (blue dye test) has a relatively high proportion of false negatives, and is not able to exclude micro inhalation.

Therefore, to improve the reliability of BSE evaluation, some authors have thought to correlate oxygen saturation with the water swallowing test. The combined detection of symptoms arising from the swallowing test and blood oxygen saturation, greatly increases the sensitivity of each test since inhalation causes oxygen desaturation (mean of 4.6%), revealed with oximetry. A decrease of more than 2% in saturation is still considered significant so the patient must be subjected to more accurate diagnostic evaluations, while a decrease of more than 5% must immediately lead to interruption of the test as inhalation has occurred. Within the Evidence Based Medicine, the oxygen desaturation test is still not fully accepted because posture, swallowing and coughing could influence the SpO_2 levels during the test. However, when the oximetry results are combined with those of the water swallowing test, the accuracy increases significantly. A recent review of the literature has shown that the combination of these two tests (oxygen desaturation during water swallowing, assessing any coughing, choking and/or voice modification) can be considered more reliable. The positivity of these tests necessitates a thorough investigation using imaging techniques such as video fluoroscopy (VFS) or Fiberoptic Endoscopic Evaluation of Swallowing (FEES).

The FESS is preferable in cooperative patients with suspected silent aspiration, and it is practical especially in cases where there is uncertain inhalation or contributing factors such as risk cough ineffectiveness. Nonetheless, both methods give reliable results in patients with LCF $\geq$ 4.

The fiberoptic endoscopy, performed during FESS, allows a good view of stagnation in the valleculae, in the pyriform sinuses, in the laryngeal vestibule. Furthermore, stimulation with the tip of the endoscope permits triggering of pharyngeal and especially laryngeal zones.

Videoendoscopic investigation allows a static and dynamic assessment of the structures of the upper respiratory and digestive tract. The static evaluation, aside giving information of the anatomical structures involved in swallowing, allows to study the function of the laryngeal sphincter and the possible detection of stagnation; dynamic evaluation by the administration of a bolus, allows to evaluate proper swallowing function.

Resuming, FEES allows:

- observation of the anatomical structures involved in the oral and pharyngeal phases of swallowing;
- the observation of motility and sensitivity of pharyngo-laryngeal structures critical for swallowing;
- observation of secretions and their management;
- direct assessment of swallowing when possible of different texture food (liquid, semi-liquid, semi-solid, solid and mixed);
- observation of the effects of therapeutic manoeuvres, used to improve swallowing.

The creation of an appropriate diagnostic-therapeutic path, through the application of dysphagia scales, such as the Swallowing Rating Scale (with a score between 1 and 7), and the functional outcome measurement, such as the Dysphagia Outcome and Severity Scale of O'Neil(DOSS), are necessary to improve the functional prognosis of the patients. A proper evaluation leads to an appropriate treatment and planning of a therapeutic path to follow (weaning from aids such as PEG, the SNG and the tracheostomy tube), especially for patients with nasogastric tube, which is not very suitable for dysphagic subjects, and that, according to the SINPE (Italian Society of Parenteral and Enteral Nutrition) guidelines, should not reside for more than four weeks.

To improve the functional prognosis of patients and to decrease the number of lethal complications,such as protein-energy malnutrition (PEM), dehydration, aspiration pneumonia, pressure sores and infections, resulting in increased morbidity and mortality, a proper pathway is needed.

The logopedic evaluation of the clinical and functional alteration of swallowing, with the application of BSE and FEES, identifies the degree of dysphagia, according to the Swallowing Rating Scale (American Speech-Language- Hearing Association-ASHA), which provides a score between 1 and 7 (from 0 to 2 will usually refer to patients in VS or MCS):

Level 0: The patient cannot be evaluated.

Level 1: swallowing is not functional.

Level 2: swallowing is inconsistent or delayed, prevents an adequate nutritional intake, although swallowing is sometimes possible.

Level 3: the alteration of swallowing partially prevents nutritional intake, and requires close monitoring of the patient during feeding.

Level 4: the alteration of swallowing does not prevent nutritional intake, but supervision is required for the use of compensatory techniques.

Level 5: altered swallowing is sufficient for nutritional intake, but compensatory techniques are required and sometimes special nutrition techniques and dietary modifications.

Level 6: swallowing is functional for most of the feeding activity but sometimes there are difficulties. The meal can last longer.

Level 7: Normal swallowing.

This is followed by the activation of Rehabilitative pathways (red-yellow-green-white), and the functional outcome measurement which follows the relative nutritional approach, through the use of the DOSS-dysphagia Outcome and Severity Scale of O'Neil.

The DOSS scale identifies the following levels of deglutitory alterations, advising for a proper oral nutrition:

- Level 7: normal oral nutrition- normal level of independence, normal diet; requires no rehabilitation and no increase in time spent on the meal.
- Level 6: minor swallowing alterations; independence with limited functionality, normal diet; functional swallowing. Time spent on meal can lengthen. The patient may have a moderate delay in oral and pharyngeal swallowing or a mild stagnation, but compensates spontaneously. There is no aspiration or penetration with any food consistency.
- Level 5: mild dysphagia - remote supervision. The patientmay require a restricted consistency diet. Mild oral dyspepsia with limited chewing and/or oral retention which is, however, eliminated spontaneously. In the case of contrast liquid aspiration, strong cough stimulus is capable of completely eliminating the aspirated material. Penetration into the airways with one or more consistencies or up to the vocal cords with one consistency, but with spontaneous clearing. Pharyngeal retention with spontaneous clearing.
- Level 4: Mild to moderate dysphagia. Supervision/directions with minimal contact. Need to avoid one or two food consistencies. Oral cavity retention, which can be eliminated on command. Retention in the pharynx removed on command. Suction with a consistency of

contrast medium with little or no cough reflex or penetration into the airways at the level of the vocal cords by coughing up two consistencies of contrast medium, or penetration into the airways at the level of the vocal cords without coughing up a single consistency of contrast medium.

- Level 3: Moderate dysphasia. Total assistance and/or supervision. Need to avoid one or two types of consistencies. Moderate oral cavity retention that can be eliminated upon command. Modest retention in the pharynx eliminated after command. Penetration into the airways up to the level of the vocal cords without coughing up two or more amounts of contrast medium aspiration or two contrast medium aspirations with little or no cough reflex, or aspiration with a contrast medium consistency without cough and penetration into the airways up to the vocal cords with a consistency of contrast medium without cough.
- Level 2: Moderate to severe dysphagia. Maximum assistance. Only partial oral nutrition is possible (can tolerate only one consistency with the use of compensation postures or other swallowing techniques). Severe incontinence or retention of the bolus in the oral cavity that cannot be eliminated spontaneously or after multiple commands. Aspiration with two or more contrast medium consistencies without cough reflex and with poor voluntary cough. Or aspiration with one or more consistency of contrast medium without cough and airway penetration up to the vocal cords with one or more contrast medium consistency without cough.
- Level 1: Severe dysphagia. One cannot employ any safe oral nutrition. Severe incontinence or retention of the bolus in the oral cavity that cannot be eliminated. Conspicuous stagnation of the bolus in the oral cavity that cannot be eliminated. Silent aspiration of two or more contrast medium consistencies, ineffective voluntary cough. Incapable of swallowing.

Rehabilitation of Dysphagia

Re-education of dysphagia needs the intervention of dedicated therapists, with regard to speech therapists specialized in deglutition and the use of treatment protocols, including the use of compensatory strategies, such as postural adaptations, sensory stimulation, manipulations of volume, consistency and viscosity of food.

Rehabilitation always starts from a position of non oral feeding. The personalized form of action may, then, follow different pathways, and includes the goals that a patient can reach to restore or remedy some functions. These paths are generally present in a constant way, although they are reached at different times:

- Path I or RED is suitable for patients in VS and/or with complete dysphagia and/or with enteral feeding. For these subjects speech therapist treatment will be scheduled to facilitate the recovery of swallowing function (remedy methods) to implement stimulation of the peri-buccal and oral sensitivity (thermal-tactile-vibratory), passive neuromotor treatment and/or active orofacial structures (massage, praxis), stimulation of the swallowing reflex, exercises to restore the costo-diaphragmatic breathing with particular attention to the stimulation of cough and apnea mechanism and management of oral secretions. It also provides counselling and information to family members, monitoring the situation and in case of improvement, the most suitable route and dietitian counselling.
- Path II or YELLOW is suitable for patients in MCS and/or severe dysphagia and/or mixed feeding (enteral and oral). The logopedic treatment is to facilitate the restoration of functional swallowing with the addition of adaptive and/or restorative methods until weaning. This includes stimulation of the peri-buccal and oral sensitivity (thermal-tactile-vibratory), passive neuromotor treatment and/or active oro-facial structures (massage, praxis), exercises to restore the costo-diaphragmatic breathing with particular attention to stimulation of the cough and apnea mechanism, stimulation of the swallowing reflex, swallowing of saliva. If possible, first attempts are made toward swallowing non-hazardous food material in small quantities (ice, water gel), which are followed by tests of swallowing with other food consistencies (mousse, creams). The pathway also provides counselling and information to families on dysphagia, monitoring of the situation and, in case of improvement, the programming of the most suitable route and advice of the dietician on the adequacy of present nutrition (nutritional intake screening).

There are two other routes (GREEN and WHITE), which, however, are suitable for responding patients and/or collaborating patients with moderate or mild dysphagia and/or oral feeding and that expect oral nutrition (dietary management, postural compensation, swallowing manoeuvres, change in the food consistency, choice of the most suitable bolus volume, attention to temperature and taste).

As can be seen from a detailed description of the rehabilitative pathway, speech therapy aims mainly at restoring adequate afferent gustatory-olfactory, tactile, proprioceptive-kinestetic, and thermal properties. Recovery is of fundamental importance for the reorganization of respiratory, cough, voice, oral automatisms and of the swallowing prerequisites.

The instruments concerning treatment choices/ or preliminary interventions are promotion of careful oral cavity hygiene, co-determination of choices regarding the tracheostomy tubes, the search for adequate posture, the implementation of automation and oral swallowing, the management of oral reflexes, stimulation and passive mobilization of the bucco-lingual-facial structures. Interestingly, a new treatment option, using neuromuscular electrical stimulation (NEMS) has recently been introduced. This promising tool, combined with the traditional speech therapy treatments for swallowing, has showed good potential for significant and lasting improvement of dysphagia.

The targeted treatment for dysphagia through the use of NMES involves the application of electrodes to the skin of the head and neck in order to obtain, thanks to the pulses of electrical energy, a stimulation of theweakened or paretic muscles. This must necessarily be combined with the standard rehabilitation of swallowing and/or rheological modification of foods and fluids. Rehabilitation treatment for dysphagia, recognized as the "gold standard" by the scientific community, is a set of operations (hand postures, swallowing manoeuvres, bolus modification, sensory stimulation, muscle strengthening), which are used by healthcare professionals in the field to treat and stabilize the dysphagic patient.

The use of NMES in the treatment of dysphagia has revolutionized the way of thinking in the field of deglutition rehabilitation. Since the beginning of the method, many controversy were born. One of the main points of discussion deals with the rapid diffusion and its application in clinical routine, without in-depth clinical studies indicating its effectiveness. Moreover, the fact that the same treatment was proposed for patients suffering from a wide range of ailments and the need to train health care professionals to its use have also stirred much debate within the American scientific community.

Although this technique may contain rehabilitative potential for certain groups of patients, under certain conditions, the widespread use of this technique does not mean that it is synonymous with a universal rehabilitative and equally effective approach for dysphagia.

These considerations are confirmed by the fact that most of the authors attempted to demonstrate the effectiveness (or ineffectiveness) of the NMES system based essentially on an original protocol, as shown in proper reviews and meta-analysis.

A final thought may be that the NMES, system born with great potential in the treatment of dysphagia, like all specific medical devices, has to be managed by experienced professionals in swallowing. Interestingly, the greatest expectation on the benefits of this tool is related to its possible ability to promote or influence brain neuromodulation, thus making its use justified in many cases of oropharyngeal dysphagia.

However, in patients with severe consciousness impairment, this swallowing protocols are generally of little or no use because are often too complex to learn and maintain. Some patients, sometimes, recover oral automatisms with effective swallowing and, therefore, the possibility to introduce sufficient daily nutritional supplements.

In any case, in order to start oral feeding, the following conditions must be met:

- one or more clinical and instrumental swallowing tests must have been carried out;
- assessments must have detected effective and safe functional swallowing;
- the patient should be permanently vigilant, and vigilance must be closely monitored at every oral administration;
- it is useful to keep the tracheostomy tube in the initial treatment phase of swallowing and feeding in order to evaluate possible aspirations of foods, using suction on the patient when necessary. However, weaning from the tracheal cannula is an important objective to be achieved quickly;
- it is possible to maintain the patient in a suitable posture for the time necessary;
- trained and reliable caregivers must be present; feeding by mouth always has a degree of risk which requires special training and attention also of the family.

The majority of patients in VS and MCS reaches the possibility of partial oral nutrition, even if often using only creamy and homogeneous consistencies. The rest of the diet is completed by the contribution provided by the PEG. The difficulties that prevent a complete oral intake may be due to a compromised oral motor function and/or to persistent pathological reflexes together with serious disturbances of consciousness that slow the timing and the amount of intake. Nonetheless, in such patients, it is still necessary to promote oral feeding in every way possible, while respecting the safety policy during swallowing. Moreover, it is encouraged both for its hedonistic value and because meals are an important opportunity for relationship and communication between patients and their families, and for the prevention of secondary damage to oral mucosa, trophic muscle, teeth and temporo-mandibular joint.

In any case the removal of the PEG must, however, be carefully weighted. In fact, the delicate balance of stabilized but poorly responsive patients, suggests that maintaining the PEG gives the possibility to quickly use, when necessary, an eternal nutrition mode that may be increased or decreased as required.

The Management of the Tracheotomy Tube

The removal of the tracheal cannula is a primary goal in the management of basic functions, even if it is believed to be useful in the rehabilitation of dysphagia, especially in uncooperative patients. However, it interferes on phonatory, deglutatory and respiratory functions, and, if maintained for a long time, it may be also responsible for serious complications (tracheoesophageal fistula, tracheal stenosis, infectious episodes related to the presence of the device itself).

The rehabilitative treatment has the initial purpose of improving the effectiveness of cough, for a better toilet of bronchial secretions and for the pulmonary re-expansion so to prevent pulmonary atelectasis and reduce as much as possible the risk of infections. The different procedural steps provide a gradual reduction of supportive oxygen, breathing with open cannula, breathing with closed cannula for progressively longer times, up to the achievement of a complete respiratory/ventilatory autonomy.

The tracheal cannula protects very little from the risk of aspiration, and, furthermore, it constitutes an notable obstacle for the recovery of normal swallowing. Closing of the cannula can determine decubitus ulcers and otherwise alter the endocervical reflex (in particular with hypercontraction of the upper esophageal sphincter). For this reason, its use should be strictly limited to the time necessary for the proper situation.

The presence of persistent disturbance of consciousness does not limit the performance of manoeuvres designed to wean from the tracheostomy tube, but involves constant monitoring of respiratory parameters, cardio-circulatory and careful clinical supervision.

Thus, the progressive autonomy from the cannula must be included in the dysphagia rehabilitation program, given the need to reaccustom the patient to breathe through the natural ways, so to improve all other functions related to them (breathing, swallowing). It is emphasized that decannulation is possible in cases of VS or MCS after having assessed for the presence of effective cough and automatic swallowing. During the weaning period, a fenestrated cannula, which is closed for longer periods of time until its removal, is provided.

The removal of the cannula can be obtained if there is a good laryngo-tracheal patency and absence of dyspnea for 24/48 hours with closed cannula, the absence of "aspiration", a good recovery of expectoration capacity, absence of lung infections, absence of tracheal stenosis, existence of a sufficient swallowing mechanism.

In the presence of respiratory rate alterations, shortness of breath, agitation, tachycardia, alterations of mechanical ventilation, cyanosis, cough, shallow breathing, an assessment is useful in order to check respiratory conditions and interrupt the abortion of the tracheostomy tube procedure.

The ability to remove the tracheostomy tube is not always attainable, especially in long-term tracheotomy patients, and thus, decannulation can be difficult and must be done with great caution while keeping the patient under close observation.

Decannulation is also possible in selected VS or MCS patients, after clinical evaluation of tolerance to closure of the cannula (for progressively longer periods up to at least 48 consecutive hours) when they meet the following criteria: saturation device oxygen greater than 92% in room air, sufficient cough efficacy with reduction and/or self-management of secretions, the absence of infection and negative chest x-ray, at least partial effective swallowing, no obstruction of the upper airway, favorable conditions of nutrition.

In any case, before proceeding with decannulation, a fiberoptic evaluation accurately assess the supraglottic region and the glottic plane in order to exclude the presence of tracheal stenosis, granulomas obstruents and tracheomalacia must be carried out.

Conclusion

Artificial nutrition is an integral part of the therapeutic approach in patients affected by vegetative and minimally conscious state, ensuring the adequate nutrients for the prevention of protein and caloric deficit responsible for the highest increase in therapeutic failure, medical complications and mortality. Among medical problems, neurogenic dysphagia is of particular importance in these patients since it may lead to pulmonary aspiration, which may manifest acutely as choking or coughing, respiratory distress, wheezing, gasping or gurgling, and tachycardia, or chronically with silent aspiration, weight loss, production of excessive oral secretions and aspiration pneumonia.

Thus, preventing dysphagia and malnutrition should play a pivotal role in the management of unconscious patients so to improve, if possible, their "quality of life".

References

Artificial Nutrition

Bernat JL, Beresford HR. The controversy over artificial hydration and nutrition. *Neurology*. 2006 13;66:1618-9.

Cherney LR, Halper AS. Recovery of oral nutrition after injury in auduls. *J Head Trauma Rehabil* 1989; 4: 42, 1989.

Finch H. Nutrition and hydration for the vegetative state and minimally conscious state patient. *Neuropsychol. Rehabil*. 2005;15:537-47.

Freeman EA. Protocols for the vegetative state. *Brain Inj*. 1997 Nov;11(11):837-49.

Jeejeebhoy KN. Enteral feeding. *Curr. Opin. Gastroenterol*. 2002;18(2):209-12.

Jennett B. Thirty years of the vegetative state: clinical, ethical and legal problems. *Prog. Brain Res*. 2005;150:537-43.

Linee guida SINPE per la Nutrizione Artificiale Ospedaliera 2002.

Ruiz-Santana S, Arboleda Sánchez JA, Abilés J; Metabolism and Nutrition Working Group of the Spanish Society of Intensive Care Medicine and Coronary units. Guidelines for specialized nutritional and metabolic support in the critically-ill patient: update. Consensus SEMICYUC-SENPE: nutritional assessment. *Nutr. Hosp.* 2011;26 Suppl 2:12-5.

Scolapio JS: A review of the trends in the use of enteral and parenteral nutrition support. *J. Clin. Gastroenterol.* 2004;38:403-407.

Dysphagia

Appropriatezza e linee guida in logopedia linee guida sulla gestione del paziente disfagico adulto in foniatria e logopedia. Consensus conference Torino, 29 gennaio 2007; Raccomandazioni del comitato promotore federazione logopedisti italiani.

Bakheit AM. Management of neurogenic dysphagia. *Postgrad. Med. J.* 2001;77:694-9.

Buchholz DW. Dysphagia associated with neurological disorders. *Acta Otorhinolaryngol. Belg.* 1994;48:143-55.

Hamdy S. The diagnosis and management of adult neurogenic dysphagia. *Nurs. Times.* 2004;100:52-4.

Hughes TA, Wiles CM. Neurogenic dysphagia: the role of the neurologist. *J. NeurolNeurosurg Psychiatry.* 1998;64:569-72.

Logemann JA. Screening, diagnosis, and management of neurogenic dysphagia. *Semin. Neurol.* 1996;16:319-27.

Malandraki G, Robbins J. Dysphagia. *Handb. Clin. Neurol.* 2013;110:255-71.

Seidl RO, Nusser-Müller-Busch R, Ernst A. The influence of tracheotomy tubes on the swallowing frequency in neurogenic dysphagia. *Otolaryngol. Head Neck Surg.* 2005;132:484-6.

Wiles CM. Neurogenic dysphagia. *J. Neurol. Neurosurg. Psychiatry.* 1991;54:1037-9.

Thacheostomy

Lewarski JS. Long-term care of the patient with a tracheostomy. *Respir. Care.* 2005;50:534-7.

Zhang H, Slutsky AS. Year in review 2008: *Critical Care--respirology. Crit. Care.* 2009;13:225.

Zhu H, Das P, Brereton J, Roberson D, Shah RK. Surveillance and management practices in tracheotomy patients. *Laryngoscope.* 2012; 122:46-50.

In: Chronic Disorders of Consciousness ISBN: 978-1-62618-940-9
Editors: R. S. Calabrò and P. Bramanti

Chapter 9

Rehabilitation Paths in Severe Brain Injury

Fabrizio Denaro, Maria Zanghì, Graziano Giuffrida and Placido Bramanti
IRCCS Centro Neurolesi "Bonino-Pulejo", Messina, Italy

Introduction

Vegetative state is a clinic condition that occurs after a coma caused by an acute event (trauma, stroke, cerebral anoxia).

People suffering from severe brain injury usually receive the primary hospital treatment in intensive care units, during this period subjects are supported by health interventions such as ventilation and dressings. The acute phase may last for a few weeks or months until the subject is stable and it is possible to start a specialist and multidisciplinary rehabilitation path, which allows the patient not only to survive but also to maintain as much as possible the necessary vital functions [1]. Once the patient is stabilized it is necessary to define what kind of treatment should be submitted, where to continue rehabilitation and who will provide the services. Decisions concerning health are jointly taken by the health care team that assists the patient and family members, the real problem is where the subject can continue the treatments. The choice for families is limited to a rehabilitation Institute or to the reintegration at home that, if ensures the closeness of relatives, but may not provide the patient with the appropriate treatment to maintain a good quality of life, while a rehabilitation Institute does not allow family members to constantly assist the patient. A family that lives in an area of difficult geographic access, if decides for domiciliation, may have difficulty to support his patient for logistical, temporal and financial reasons; having to deal with everyday life that often doesn't allow to assist him. This last reflection is to understand that people in VS or MCS are part of the category of vulnerable people, for which law must provide protection and care assistance. The approach of care and rehabilitation to people with low responsiveness due to injury in brain structures is a problem of serious medical and social relevance. The VS due to severe brain injury and resulting in coma has, in its initial phase, a real need to be taken in intensive care. The inclusion of such patients in intensive care in a first phase is essential for survival but becomes a critical factor when the patient is stable and long hospitalization in these departments becomes "inappropriate" and reduces the resources for other types of patients. What is shared in the scientific community is the need of departments dedicated to these diseases that, as shown by some studies, have a lower incidence of mortality and a better prognosis than those treated in non-specialized departments.

International Models

Currently there aren't international guidelines for the management and care of patients with the disease just described. What seems to be shared, although not codified, is the need to develop a process that can allows the patient to be recovered in the time required, ensuring the best quality of life for his family and a better use of resources by the national health services. In this analysis can be useful to describe some international models of reference, the Italian model which will be treated separately and the models of three different countries and continents: France, United States and Australia.

In France patients in vegetative state are treated in dedicated care units; this model has the idea of "continuous assistance" and include a large number of subjects in VS. These units are small "houses" that can accommodate form 6 to 8 people, their position ensure easy access in terms of proximity to family and public transport and having as reference a about 300,000 persons per unit [2]. Admission to such structures doesn't consider a defined period and, based on what indicated by the French Ministry of Health, care units are located in health facilities with a multidisciplinary team and that meet the specific requirements set out in the Ministerial Circulars.

After the discharge the patient has two options: the hospitalization in a medical centre or the domiciliation. The latter hypothesis is possible only when the family has been adequately informed about the difficulties and the physical and psychological implications that the management of a patient in this condition requires. This is possible only through a close correlation between the health and social personnel and the family.

Another model under consideration in this work is the Australian that, in the immediate aftermath of the trauma provides the admission to the hospital for patients in an acute but stabilized state (not before the 4th week after the event) [3]. Then patients are transported to a rehabilitation facility, a nursing home, a community of support and care or at home. Some people may choose only one of these options, some may choose several, others may move from acute care to home care for a monitoring and stabilization period and to test the necessary rehabilitative activity. In Australia the facilities for management and care of subjects in VS are: *Residential care homes,* that are structures not specifically appropriate for this type of patients, but with the presence of physiotherapists and speech therapists; *Community houses,* that are communities with about 4/5 people with similar disabilities, for whom assistance is available 24 hours a day and that develop customized programs of rehabilitation. Community care can provide an adequate care for people in VS and MCS if it is provided with the necessary equipment and expertise to ensure an appropriate health care and medical rehabilitation; *Care at home,* the traditional home care performed by the patient's family [4].

In the U.S.A., Brain Trauma Foundation's QI TBI Program is a cooperative consortium of trauma centers dedicated to improving severe TBI care in the acute care setting. The QI TBI Program offers a step-by-step approach for incorporating quality and process improvements into the care delivery model for TBI. It's designed to help trauma centers measure and monitor their performance on key clinical indicators based on established evidence-based guidelines. This can improve patient safety and quality care, as well as reduce operating costs. Severely brain injured patients spend the most critical days of their lives in trauma centers under the care of a variety of specialists who are key to achieving improvements in outcome. Understanding the process of how TBI care is given by these specialists through national surveys, allows BTF to bridge gaps in knowledge and improve patient care. The implementation of TBI Guidelines produces improved outcomes beginning with the pre-hospital phase and extending throughout long-term care.

Pre-hospital management of TBI includes correctly identifying the TBI, optimal treatment in the ambulance, and direct transfer to a TBI trauma center.

In-hospital guidelines application leads to: a reduction in ICU days & costs, decrease in deaths & disabilities and improved neurological outcome upon discharge. In addition, it's projected that if all US trauma centers adopt the BTF guidelines, there would be a cost savings. Medical personnel and families are educated on long-term patient prognosis.

Half of those who die from TBI do so within the first 2 hours of injury. It is now known, however, that all neurological damage does not occur at the moment of impact (primary injury), but rather evolves over the ensuing minutes, hours, and days. This secondary brain injury can result in increased mortality and disability. Consequently, the early and appropriate management of TBI is critical to the survival of these patients. Prehospital assessment and treatment is a critical link in providing appropriate care. Treatment begins in the field and continues during transport by EMS providers. However, a growing trend to offer additional training to the EMT-Basic now permits some of them to perform some invasive procedures. The EMT-Paramedic, who has the highest level of EMT training, is allowed to perform advanced patient assessment. In many countries, physicians in ambulances or helicopters respond to the call and care for the patient in the prehospital setting. Guidelines are not intended to supplant physician judgment with respect to particular patients or special clinical situations and are not a substitute for physician-patient consultation. The adhesion to these guidelines is voluntary and must be applied by a doctor, according to the specific situation of the patient [5].

Italian Model and Meaningful Experiences

In Italy there isn't a specific legislation for Vegetative State, even if we find mention in the draft law on "Organic Testament" which is under discussion in parliament. In recent years two ministerial committees have been established, the so-called Di Virgilio in 2005 and the Roccella in 2009 [6], to discuss and highlight the main issues related to the subjects in a Vegetative State. The work of the two commissions has given rise to the "Guidelines for the care of people in Vegetative State and Minimally Conscious State" approved by the State-Regions Conference on May 5, 2011 [7]. The definition of the guidelines derives from the need to meet the difficulties in the management of a patient in VS o MCS, increase when poor families are involved. The approach of care and rehabilitation of these people is a medical and social problem of great importance, as their life expectancy is constantly increasing in all industrialized countries, and for the advancement of knowledge and techniques in the field of emergency medicine, for progress in rehabilitation strategies and for the improvement of the quality of long term. The health care is heterogeneous, fragmented from region to region and often the care staff is unprepared to handle this type of patients; the result is an inadequate rehabilitation. Often, also the process of domiciliation is inappropriate. The domiciliation begins at the time of the discharge from the intensive care unit, once the vital functions of the patient are stabilized. In this phase the patient should be admitted in structures of high rehabilitative specialization, the lack of specialized structures makes difficult this path. Moreover, some structures that declare themselves organized to provide assistance to people in SV are not accredited by the NHS, which results in economic participation of the family expenditure to be incurred for hospitalization, provided that, if it were a contribution to support the family, this varies depending on the ASL office. There are also families who decide to turn to healthcare facilities across the border and are forced to bear the economic burden that is particularly heavy, with the result that the same community must still cope with the extra economic weight. Some families decide for solutions not offered or indicated directly by the structures (from other regions or abroad), that in the "paper" seem to be able to provide proper and adequate care to their relative [8]. To better understand the real problem it's necessary to return some data. A national project, funded by the National Center for the Prevention of Disease Control, entitled" Functioning and Disability in the vegetative and minimally conscious" [9], conducted a study on caregivers. In the project have been examined the cases of 602 patients (566 adults and 36 children patients), of which most patients in long-term care facilities (25 facilities), in other rehabilitation facilities (38 facilities) and other at home. Approximately 69% of patients are collected in SV and 31% in SMC. The data have been provided by 485 caregivers, of which 70% are female. The average age is 56 years. 26% are retired and 22% at home. 67% of the sample of caregivers' patients with SV SMC report having reduced all the leisure activities than before the event of their relative. 39% declare to assist three hours per day, while 24% between 4 and 6 hours daily. The 80% of caregivers instead declares continues to provide assistance 24 hours on 24.

Finally, CENSIS data of 1999, which highlights the social exclusion that can result for caregivers. These data show that about 65% of caregivers was forced to change or leave their jobs in order to ensure patient care. The daily commitment required to assist a patient in VS is physical, psychological and economic. From a cost analysis stands out for families an economic burden of about 3.000-3500 euro a month [10], it becomes a burden to discriminate against low-income families, and makes domiciliation a privilege only for wealthy families. in order to ensure an adequate and rehabilitative care pathway had been established by the so-called Commission Di Virgilio, the Special Unit Home Permanent (OSS) for subjects in a vegetative state or Minimally Conscious, but were made patchy in the territory Italian, often not integrated with the realities of the territory. This type of approach is taken with the commission Roccella and then shared in the Guidelines for the care of persons in Vegetative State and Minimally Conscious State. The proposal is for a better flow of patients in order to enable a more comprehensive redevelopment of care pathways for patients with severe brain injury acquired in a strong territorial characterized by universality, appropriateness, timeliness and progressive care.

The clinical phases are: critical / acute/ subacute / post-acute. For each phase is essential to define an appropriate therapeutic and rehabilitative intervention, that support a strong cooperation between different structures for the different phases of the patient's recovery. To this end, it should be favored the establishment of an integrated network, a path of social welfare to accompany patient's clinical onset in a vegetative state and accompany him to his return to the family. It is also important to establish a national network of reference (Figure 1). The whole system must be governed by interdisciplinary approach, professional integration structural subsidiarity and by providing in each phase, the response to any diagnostic therapeutic needs of the patient.

In this context, the Ministry of Health has funded line 3 of the priority objectives of the National Health Plan aimed at promoting their organized care's models for patients in chronic vegetative state; intervention line 3.1 units of special care pathways to permanent accommodation (Suap) for subjects in Sv or SMC. Currently in Italy we have a fragmented framework that has seen some regions anticipate the possibility of social care interventions to these patients in line with the provisions of the documents Di Virgilio and Roccella, in this area of absolute value are regulations approved in Sicily - *DGR 21.2.2007, n. 24 - "Regional Network for assistance to persons in a vegetative state and minimal knowledge" and the Decree of 21 December 2011 "Identification of Special units receiving permanent" - OSS, Lombardy*

DGR 19.12.2007, n.VIII/6220, province Trento-DGP 24.08.2007, 174- "Patients in a vegetative state or minimal responsiveness: the first guidelines for the construction of Network-hospital care", Puglia regional regulation 2 November 2011, n. 24 "Model rehabilitative care and to take charge of the subjects in Vegetative State and Minimally Conscious State - territorial requirements and minimum structural, technological and organizational requirements for the authorization and accreditation of rehabilitation facilities outside hospital_ Awakening Centre".

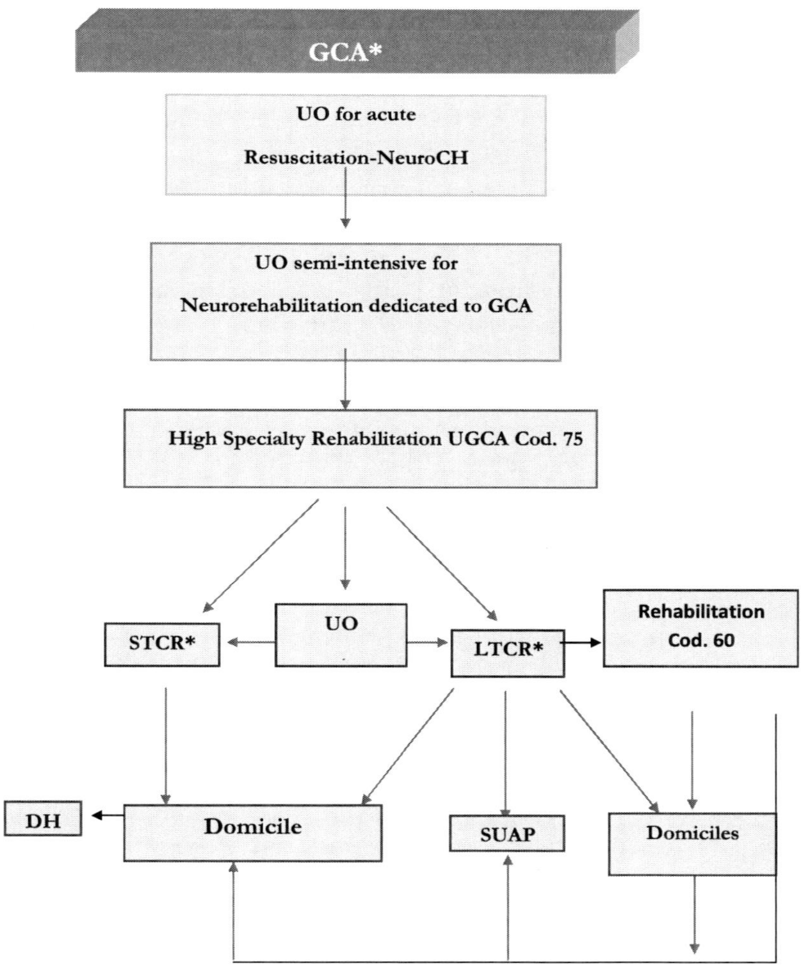

* STCR: Short Term Comprehensive (hospitalization measured in weeks).
*GCA: Severe Brain Injury Acquired.

Example of great interest is that of Sicily, where in 2007 [11] were upheld the directions of the Di Virgilio Committee and was approved a decree and initiating the establishment of a regional network for the vegetative and minimally conscious subjects. Network was reorganized with a new decree in 2011. If in 2007 they were designated 20-bed unit for Awakening and 80 OSS, distributed at the San Raffaele Foundation, the PO Giglio of Cefalù, the IRCCS Bonino-Pulejo of Messina, the PO of Mistretta and the P.O. Caltagirone, of which 10 units of awakening and 20 OSS at the IRCCS Bonino-Pulejo of Messina, with the decree of 2011 of Health Regional Department [12] are provided a total of 155 beds not part of a hospital network, with OSS centre for each province of Sicily. In this context plays an important role the Institute for Research, Hospitalization and Scientific Care Center Neurolesi "Bonino Pulejo" of Messina, institution to which were assigned, in 2007, 20 beds for the OSS and 10 beds for the Awakening Unit. The Institute in recent years has been characterized by high specialization in assisting individuals in SV and SMC and by research carried out in this type of subject. It stood out as a true HUB of the regional network for Vegetative States. This favored a grant from the Sicily Region, for the priority objectives of the National Health Plan for the year 2011, that enable the Institute to test a pilot project for the creation of a social – assistance - rehabilitation training for people in SV and SMC, and specifically the construction of a house accommodation which allows through a path of self-management the gradual reintegration at home of the subject, which allows to interact directly with the IRCCS. The house provides accommodation in addition to a patient's bed for a family member who can then assist 24h. The peculiarity of this project, than the path provided by the ministerial committees, is the possibility of using telemedicine.

In the Italian context, has certainly an important place the experience developed around the House of Awakening Luca De Nigris of Bologna.

The House of Awakening Luca De Nigris is a public hospital that has the main objective of providing a service of assistance and rehabilitation to patients who are in a vegetative or post-vegetative state in post-acute phase with still potential for change. From the point of view of health services, the center is part of a particular segment of the path of hospital care that is preparatory to clearance. In this sense, the House of Awakening is an ideal "bridge" between the hospital and the home of the patient, the House of Awakening Luca De Nigris in addition to be a health intensive care for severely disabled persons and completely dependent not serviceable at home and requiring ongoing rehabilitation treatments, will also be a research center for the development of clinical procedures, care, treatment and assessment for the vegetative and immediately post vegetative patient. The objectives of the House go beyond those covered by traditional hospitals. The experimental aspect of the therapeutic model pursued by the House of Awakening is in fact to consider how patients, not only the affected person from the coma, but his primary system of relations, typically detectable in his family. In this model is fundamental the integration between: health (doctor, nurses, health staff, therapists), extra-care (volunteers of "The friends of Luke ") and families. In the phase of the outcome, people with disabilities and their families need to rely on a network of opportunities and local services able to ensure their return to their family and social context to which they belong. Going back to the original plan of the House of Awakening, this is based on treatment and care interventions, carried out daily in the hospital department of resuscitation in a physical context characterized only by patients awakened from a coma. The model of care emphasizes the central role of the family and the possibility of continuous coexistence of family with an active and aware role, maintaining relationships, rhythms and rituals of everyday life. Operational cooperation between the Local Health Unit of Bologna and the Association "Friends of Luca" allows the presence of various professionals, not just health care, and organized volunteers that promote flexibility in the rehabilitation difficult to achieve in a traditional hospital structure. Inside the structure is also allocated the Research Center for Research on Coma (director Fulvio De Nigris) as part of the project, the need is to create cores of research closely to combine care and research in order to achieve a virtuous circle: the power assistance research and the latter has a transfer on clinical practice that allows the improvement. The integration between the different elements of service doesn't simply mean that each of these individual elements are coordinated, the concept of service to which the House of Reawakening refers to is that of a set of elements each of which is essential for the overall system as a pillar of one roof: Furthermore, each of the elements has positive externalities on others; without one of these the other elements is considerably reduced effectiveness. The innovative element in the therapeutic care followed by the House of Awakening is the unique systemic approach to patient care. The reference unit for the service staff is not only the subject of patient care, but also involves his family. Between the family and the staff there is a continuous interaction. The model of service delivery that results is multiprofessional, it's not a division of labor or skills based on a rigid hierarchy, but a variety of professions that complement each other. The project has the main objective to "rehabilitate", meaning by this term, not the complete cure of a disease, as the implementation of all the methods and strategies needed to achieve the reintegration of the person in the social body, at the highest for it to reach self-sufficiency level of physical and cognitive. The Regional Regulation 2 November 2011, n. 24 of the Puglia Region [13] establishes "the organizational model of the path of the subject with severe acquired brain injury (GCA), discharge from the hospital to the home," in line with what established by the Guidelines "to assist people in a vegetative state and minimally conscious state" These structures are called " Extra Hospital Rehabilitation Structures Center Awakening" (SRE Center Awakenings), inside the care process develops in intensity levels of care and rehabilitation gradual, which correspond to the levels provided in the outpatient LR 23/2008. The Awakening is composed by the Intensive Rehabilitation Unit, the Extensive Residential and semi-residential Unit and the Special Unit of Permanent Home. They share the general structural and engineering requirements as Regional Regulation 3/05, section D. As part of the territory the number of such structures must be limited in such a way as to concentrate them at an adequate number of patients, using the available resources in proportion to their costs. To this end, each unit must be organized in modules of 10 beds.

In the phase of the discharge from the hospital, the patient may access a core of dedicated care: the OSS. The SUAP_GCA are residential facilities that provide a level of care for people in VS and MCS equivalent to palliative (palliative facilities for people with an outcome of GCA) and are connected structurally and/or functionally to the Units of Intensive and Extensive rehabilitation. These are facilities for people in VS or MCS and where voluntary associations and families receive appropriate prior training, that help to recreate conditions around the patient during care processes that might facilitate maintenance of optimal conditions for rehabilitation achieved.

Conclusion

From the above description it is clear that, although there are no common guidelines on the management of health and social subjects in Sv and SMC, it's possible to identify internationally agreed procedures. In health realities of the countries analyzed, the problem is that the subject on MCS or VS exceeded the acute phase is considered stabilized and is no longer subject to medical management but is directed towards a social management, being no longer considered "patient" but "chronic disability". The economic and social cost that the management of these patients involves is high and, for the health service of some countries, the costs for maintaining them in health facilities can be inadequate. Through national legislation and guidelines is then planned the construction of social structures that guarantee optimal patient management, the continuation of the rehabilitation and the gradual reintegration of the subject at home. These structures are divided into categories according to the level of care that the patient needs and allow the formation of family members and caregivers in patient's home management, psychologically supporting them. Another consideration is that related to the costs that families face in the rehabilitation and management of patients; in Italy are estimated between 3,000 and 3,500 Euros per month while social costs that go beyond the weight that the family must endure in everyday life are not quantifiable. It's important to define international guidelines for the health and social management of subjects in VS and MCS, in order to share experiences, studies and social aspects that go beyond the health protocols and tha can help the development of national networks for a better management and taking care of this kind of patients, in line with the budgetary needs of individual states.

References

[1] Multi-Society Task Force on PVS. Medical aspects of the persistent vegetative state. Parts I and II. *N. Engl. J. Med.* 1994;330:1499-1508, 1572-1579.

[2] V. Saou^t a,*, M.-P. Ombredane b, J.-M. Mouillie c, C. Marteau d, J.-F. Mathe´ e, I. Richard a Patients in a permanent vegetative state or minimally conscious state in the Maine-et-Loire county of France: A cross-sectional, descriptive study. *Annals of Physical an Rehabilitation Medicine* 53 (2010) 96–104.

[3] National Health and Medical Research Council, Australian Government National healt and Medical Research Council. Post-Coma Unresponsiveness and Minimally Responsive State A guide for families and carers of people with profound brain damage. *Working to Builda Healthy Australia* 2008.

[4] National Health and Medical Research Council, Australian Government National healt and Medical Research Council. Ethical guidelines for the care of people in post-coma unresponsiveness (vegetative state) or a minimally responsive state. *Working to Builda Healthy Australia*. 2008.

[5] Brain trauma foundation, Guidelines for the management of severe traumatic brain injury 3rd edition. *Journal of neurotrama*, vol. 24 supl.1 2007.

[6] Working Group Ministry of Health, Vegetative State And Minimum Consciousness" Epidemiology, evidence and models of care". *Final Document* 4 December 2009.

[7] Joint Conference State-Regions. *Guidelines for the care of people in a vegetative and minimally conscious state*. 5 may 2011.

[8] Bergami M, House of awakenings Luca De Nigris: A case study of the organizational and management model. Alma Mater Studiorum Bologna University.

[9] Regional Health and Social Agency Region Emilia Romagna. The integrated care pathway in patients with severe brain injury acquired. Acute and post-acute. Comparative analysis of regional organizational models.

[10] Ministry of Health, White Paper on Vegetative and Minimally Conscious States, the point of view of the association representing the family. *Seminar on Permanent Vegetative States*, May 2008.

[11] DGR 21.2.2007, n. 24 - "Regional network for the assistance to subjects in a vegetative and minimal consciousness state. "*Official Journal of the Sicilian Region*, n. 22, may 2007.

[12] Sicilian Region Decree of 21 December 2011 "Identification of Special units for permanent receiving." *Official Journal of the Sicilian Region* 27 January 2012.

[13] Puglia regional Regulation 2 november 2011, n. 24 " Rehabilitative and care model to take charge of the subjects in Vegetative State and Minimally Conscious State - Territorial Requirements and minimum structural, technological and organizational ones for the authorization and accreditation of extrahospital rehabilitation facilities. Awakenings Centre. *Official Bulletin of the Region of Puglia* 10 August 2011.

In: Chronic Disorders of Consciousness ISBN: 978-1-62618-940-9
Editors: R. S. Calabrò and P. Bramanti

Chapter 10

Legal and Bioethical Issues in Chronic Consciousness Disorders

Rocco Salvatore Calabrò and Vincenzo Barone
IRCCS Centro Neurolesi Bonino-Pulejo,
Messina, Italy

Abstract

Management of patients in unconscious state inevitably raises ethical and legal questions about the appropriate degree of life-sustaining treatment.

Most ethical analyses conclude that the decision to treat aggressively or passively should be guided by reliable information about how the patient would wish to be treated in this condition, in accordance with the principles of patient-centred medicine.

Once it is reliably established how the patient wished to be treated, it is within the standards of accepted medical practice to treat patients in vegetative state and minimally conscious state aggressively to permit them to live as long as possible, or to withhold life sustaining therapy, including artificial hydration and nutrition, to allow them to die. Disputes among family members over the nature of the patient's true wishes can be mediated by a hospital ethics committee and, if unresolved, referred to court for judicial review.

The patient's prognosis for functional recovery is the main element in identifying the appropriate level of treatment, and should be stated clearly.

Introduction

Disorders of consciousness (DOC) include coma, vegetative state (VS) and minimally conscious state (MCS), as well as different degrees of impaired and diminished consciousness. DOC may occur at all stages of life, from infancy to old age, since they can result from dementia, congenital disorders, traumatic and anoxic brain injuries and illness.

Within the past decade, the MCS has been recognised as a distinct disorder, leading to the realisation that significant numbers of patients previously thought to be permanently unconscious might instead be conscious, although they exhibit little behavioural evidence of conscious awareness. Evidence of hidden consciousness in patients previously thought to be permanently vegetative has been found in both neuroimaging studies and refined behavioural tests.

The rate of misdiagnosis of DOC is estimated to be greater than 40%, and misdiagnosis tends to run in one direction: patients are diagnosed as unconscious when they are actually minimally conscious. It is exceptionally challenging to distinguish clinically between the VS and the MCS [1-2].

Ethical Issues in Diagnosis and Prognosis

First of all, in order to provide the appropriate treatment, patients affected by DOC should be properly diagnosed [3]. Indeed, the diagnosis of a patient's degree of awareness is challenging and mainly based on the behavioural repertoire spontaneously demonstrated by/or induced in the patient. Although diagnostic studies, including functional MRI (fMRI), evoked potentials, electroencephalogram, may be of some help in making a diagnosis, they cannot be longer considered as specific markers of awareness [4].

However, functional neuroimaging technologies, such as fMRI, raise the possibility of refining our understanding of human cognition and awareness in severely brain-injured patients. This is a promising development in terms of enhancing the ability to provide accurate prognostic information, and possibly treatment options, to patients with DOC. The increasing use of these technologies will, however, inevitably present a new set of ethical challenges as we move towards a new understanding of complex states of consciousness.

Indeed, as Fins observed, *"Confusion will be the price of incremental progress"* [5]. Given the uncertainty of clinical diagnosis of patients with severe brain injury, there undeniably will be pressure from families to use such advanced technologies before they are appropriately refined or understood. Research on a lager scale is necessary; however, before functional neuroimaging can ethically be integrated into clinical practice, it should be recommended not creating false hopes among relatives of PVS patients by explaining that neuroimaging results remain, at present, investigational [6].

On the other hand, ethical concerns should be raised about the participation of DOC patients in functional neuroimaging studies that may require invasive procedures and have no therapeutic purpose [7]. By definition, DOC patients are very vulnerable, deserve special procedural protections, and cannot give informed consent, which is obtained from family or legal representatives according to the country. Thus, ethical committee decisions should balance between medical research and protection of the patients and their relatives, who could be exposed to a relevant psychological burden [8]. Every effort should be made to protect the wishes and directives of DOC patients legally. Moreover, research on persistent VS (PVS) and MCS is attracting public attention because of its more sensational features, including a) the persistence of some mental activity in rare cases; b) the exceedingly rare late recovery; c) the possibility that some patients are not really in PVS, but in a complete Locked-in Syndrome, recognizable only by means of an expert behavioural examination supported by neurophysiological and fMRI investigations.

If the proper diagnosis of DOC patients is difficult, it can be even harder to accurately assess their prognosis [1, 9]. DOC are syndromes that may be caused by many and different pathologies, such as brain injury and infections, with clinical pictures ranging from coma to VS, VS to MCS.

To this end, guidelines by the Multi-Society Task Force on PVS [10] are surely of some help. Indeed, VS is called *persistent* if it lasts more than 1 month, and *permanent* when irreversibility can be established with a high degree of certainty. Thus, a recovery from post-traumatic PVS after 12 months is believed to be really unlikely and recovery from non-traumatic PVS after 3 months extraordinarily rare. However, clinicians should be aware of several specific problems in delivering prognosis (that may lead to the decision to stop life support) in DOC patients, also considering the many personal bias [11-12].

Moreover, the impact of physicians bias (including their beliefs on the quality/dignity of life of patients affected by VS, which someone may consider "as alive as the plants") on medical decision made by families of the DOC patient should be taken into account.

This process ideally should involve a balance between undue pessimism regarding prognosis and provision of false hope, remaining vigilant about the possibility of personal bias, and consciously presenting the available information in as objective a manner as possible. Although it is reasonably difficult for surrogates to make decisions in the face of prognostic uncertainty, ethical care involves being direct and honest about the limitations of available data regarding prognosis and appropriately humble about what can actually be known about a patient's true state of consciousness and the possibility of recovery over time [13].

To Treat or not to Treat: The Right to Die!

Once an appropriate diagnosis and prognosis for a DOC patient are evidently made, important decisions concerning their proper management, including the beginning and continuation of life support technologies (LST), artificial nutrition and hydration (ANH), tracheostomy, and gastrostomy [12-13], must be made. Guidelines for management of patients in VS - published by the American Medical Association, the American Academy of Neurology, the American Neurological Association, the British Medical Association, the Royal College of Physicians, the Italian Neurological Society, and the World Medical Association Surveys of American and British Neurologists and Neurosurgeons - generally show support for the aforementioned positions, though they also reveal the expected diversity of opinion on questions of diagnosis and treatment [8].

Precedent-setting judicial rulings have established legal standards of decision-making for patients in VS and MCS. For example, the US Supreme Court ruled in *Cruzan* that all citizens have a constitutionally protected right to refuse LTS, including AHN, and that this right can be exercised by a surrogate decision-maker on the patient's behalf. In the UK, the Family Division of the High Court in London ruled in *Bland* that ANH could be removed from a patient in PVS.

Nevertheless, despite the abovementioned guidelines, decision making regarding the end of life is challenge [8;14]. To this end, it could be important if DOC patients had executed lawful advance directives (i.e. of a legal health care surrogate and a living will) before their illness.

However, few patients have living wills that, moreover, are too general to be of particular use, and often they cover only those situations in which a person has been determined to be terminally ill. Indeed, many people do not provide a living will for situations in which a person's degree of consciousness is profoundly affected but whose life expectancy is not necessarily profoundly abbreviated [3]. Thus, given the wide spectrum of potential disability and illness that an individual can face, it is impossible for any written directive to outline preferences in every possible scenario. Moreover, there is a well-studied disconnection between the theoretic preferences that a person might express and a person's actual preferences once faced with the contemplated situation.

Another important issue rises when a patient has no legally designated a surrogate: in this case state law governs may and should act as a surrogate [13]. When present, the surrogate is obligated, as well as he can, to make the same decisions for the patient that the patient would have made himself. If the patient has specifically expressed his wishes, the surrogate is obligated to follow them, even if they are not coherent with the decision the surrogate would have made for himself under the same situations [15-17]. Moreover, particular attention should be paid to those surrogates who are going to withdraw treatment from vulnerable patients out of self-interest, especially in the presence of familiar contrasts.

On the other hand, the continuation of LST of a PVS, who might have found living in such a state insupportable, might cause important harms, including burdens on caregivers, dignity and quality of life of the patient and "excessive" and unwanted medical treatment [18]. Indeed, it is well stated by multiple commission reports that physicians are not required to provide aggressive medical treatments when they are ineffective. It is generally accepted that physicians are either ethically obligated to operate the best interests of patients under their care, or charged with overseeing the use of finite medical resources, although it remains controversial to what extent physicians should consider just allocation of resources in any given treatment decision [19]. When physicians are encouraged to make individual treatment decisions with a focus on the proper use of resources, conflicts between the interests of individual patients and societal interests may arise. Thus, decisions about the rational distribution of health care resources should be made at the policy and societal level, separate from individual treatment decisions [3].

A particular and tricky ethical question relates to the mode of death. Stopping food and water inevitably leads to death within 14 days from dehydration. Conscious people suffer greatly if they die from dehydration. Moreover, this mode of death precludes the use of any organs for transplantation, which may run counter to the patient's known wishes. To this end, it would be possible to let the patient die more directly, thus reducing the stress on and distress of relatives and healthcare staff and allowing the organs to be used, which could satisfy at least some of the patient's previous autonomy.

Special Consideration on MCS

MCS patients may also be worse off than VS patients because of the erosion of their legal rights, and, in particular, the right to die. Because persons in a MCS are conscious and not terminally ill, their legal status is as complex and difficult to discern as their neurological status. Because they do not possess the capacity to direct their own care and treatment, they are left in legal limbo when it comes to exercising the right to die. Case law on MCS patients lacking advance directives has taken a very conservative approach to ending life-sustaining treatment, recognising that these patients are in a unique dilemma. We cannot know if they wish to live or die, and, indeed, we do not even know if they have wishes or desires, and so the default position is to maintain life. Determining when existence is no longer subjectively valuable for an individual with severely limited capacity to communicate is a vexing problem, especially for that person who emerges from unconsciousness, but fails to recover sufficiently to make his desires and interests unambiguously clear. Assuming that all persons have the same right to die, individuals in the MCS lack the means to exercise that right effectively. If the right to die is to be taken seriously, it is imperative that surrogates remain empowered to act on behalf of these most vulnerable persons, to make decisions that accord with their interests, values, and (when possible) previously expressed wishes [20].

Eluana Englaro: The "Breaking" Case and the Italian Medical Perspective

Ms. Englaro was a 36-year-old woman who was in a PVS for a period of 17 years, sustained by ANH. When her father proposed the removal of the feeding tube because Eluana ''had clearly expressed the wish to die in case of an accident that left her in a vegetative state,'' he was locked in a public effort against Catholic sensibilities and Italian judges, who had rejected his request three times. It must be pointed out that euthanasia, in the sense of ending the life of an individual who is suffering from a terminal illness or an incurable condition, is prohibited in Italy. Nevertheless, there is a general consensus in condemning ''heroic measures'' in some situations when therapy has no more effect on a patient's illness. Thus, in such cases overmedication is considered to be unethical for the patient, even though no legal definition exists on this subject [21; 13]. Patient's advance written directives, specifying how they would like to be treated should they become incompetent, are not recognized by Italian law, and relatives have no legal right to make decisions in such cases. Europe is deeply split over the debate on end-life directives; euthanasia and all forms of assisted suicide are indeed forbidden in most European countries. The Netherlands, Belgium, and Luxembourg have legalized euthanasia, assisted suicide is not illegal in Switzerland, and new medical guidelines allow doctors a sort of passive euthanasia in Sweden [22-25]. Euthanasia and physician assisted suicide is mostly illegal in the USA [26]. In all countries there is a wide range of public opinion about euthanasia and the right to die; only a few nations, including Italy, currently lack living will legislation.

In November 2008, the Italian Supreme Court accepted Mr. Englaro's latest request for withdrawal of ANH of his daughter Eluana. The court stated the following fundamental principles: the first dealt with the scientific evidence that PVS for 17 years had irreversibly damaged her cognitive functions; in the second point the court stated that there was clear and convincing evidence that Ms. Englaro would have refused ANH if she could speak for herself under such circumstances; finally, the principle that active euthanasia is prohibited in Italy was also affirmed.

Nevertheless, refusal of ANH, which is compared to medical treatment, is a right enacted by the Italian Constitution to protect a patient's autonomy regarding health and, as in this case, life. Following this groundbreaking decision by the court, public opinion polls suggested that almost half of the population was in favour of allowing Eluana to die. However, there were also those who believed in Catholic values regarding life, and they did not consider ANH to be a form of medical treatment on the grounds that it addressed no pathology [27].

A recent national survey of Italian physicians' attitudes towards end-of-life decisions following the death of Eluana Englaro represented a good opportunity to focus on some very important topic such as euthanasia, assisted nutrition and hydration, PVS, and patient's advance written directive regarding end-of-life care [28]. Indeed, the ethical principle of respect for patient autonomy and the right to refuse medical treatment for any reason, even though such a decision may untimely lead to their death, necessarily involves physicians, whose role is to evaluate patients' best interest in such situation. Respect for patients' consent of withholding therapy or their request for euthanasia creates an ethical, moral, and legal dilemma between providing medical care and respect for patients' wishes. On these bases, several studies have surveyed physician attitude towards end-of-life decision-making [22-26]. The common findings in most Western countries are that end-of-life care encompasses many non-medical aspects, such as religious beliefs, cultural expectations, legal status, and personal values. Respect for patient autonomy and dignity is commonplace, at least when the patient is conscious; furthermore, therapeutic obstinacy is also disapproved of in all countries, but it is rarely written into law. Generally speaking, Italian physicians may adopt different approaches regarding end-of life, which are influenced by moral and ethical background. Regarding this first-ever Italian survey, nearly 70% of responding doctors were aware of PVS diagnostic criteria; most of them (61%) considered tube feeding to be a medical therapy, and 66% of respondents believed that withdrawal of ANH might be appropriate depending on the patient's wishes; moreover, even though 50% of doctors surveyed were not in favour of euthanasia, a significant percentage (42%) did approve of it, while 8% of this sample was uncertain [28].

The Catholic Perspective: Pope John Paul Declaration

On March 20, 2004, Pope John Paul II issued a statement to the International Conference on "Life Sustaining Treatments and Vegetative State: Scientific and Ethical Dilemmas" on the provision of food and water to patients in a "vegetative state." In the past 25 years, there has been a great deal of controversy about the level of awareness of those in this state, their ability to experience pain, and their cognitive capabilities. There has also been extraordinary controversy about the moral requirements to treat and sustain individuals in this state. At this conference, it became clear that contemporary medicine knew very little of the abilities, experiences and cognitive abilities of those in this state, and a number of speakers made it clear that it is not at all certain that patients in coma, persistent unconsciousness or "vegetative state" are beyond thought, pain or experience of the external stimuli. It is because of this uncertainty about their condition that Pope John Paul II issued this statement. In it, he wanted to give every moral protection to patients in a "vegetative state" and give no moral justification to those who would end their lives by an act of deliberate omission. In his statement, the Holy Father declared that the provision of AHN was not a medical treatment, but was "basic" or "normal care" and that there was a "moral obligation" to provide it to the persistently unconscious. He denied that tube feeding could be administered to patients according to the traditional standards of "ordinary and extraordinary" care and treatment, and called these measures "basic health care," and a "natural means of preserving life." The Pope's statement sought to prevent Catholics and Catholic health care facilities from becoming involved in euthanasia by omission that has become a fast-growing threat.

Like Papal statements of previous decades on organ transplantation, abortion of ectopic pregnancies, and on pain relief for the dying, this allocution on feeding those in a "vegetative" state is authoritative, binding on Catholics and Catholic health care institutions because it sets a moral standard for Catholic health care and protects innocent human life. Seeing the vulnerability of persons who are persistently unconscious, Pope John Paul has drawn a clear line in the sand, stating that the deliberate withholding or removal of food and water from these patients cannot be done without an intention to bring death upon them; thus, death resulting from this kind of decision is inhumane, cruel and an affront to their dignity [29].

Sacrality vs Dignity of Life

Patient care in the last stages of life presents new approaches to the patient-doctor relationship. In the last years, the way to go through the final part of life has changed in its cultural, scientific, technological and even spiritual areas. For instance, this unknown stage is different in the city and the country. In some rural societies we can still witness the medieval "tamed death"; death doesn't come as a surprise and we can even be prepared for it. In the urban context, death is denied, forbidden, intervened! On the other hand, the development of medicine together with science and technology have led with all its progress to the medication of several moments of life, particularly those of the beginning and the end of life, even to the point of what we call "therapeutic obstinacy".

Marc Augé stated: "To die without dignity is to die alone, abandoned, in an inhospitable and anonymous place, in a non-place - as stated by. To die without dignity means to die suffering needlessly or to die tied up to a technical gadget that becomes the sovereign of my last days. To die without dignity also means to die in isolation, surrounded by insensitive people, soulless specialists, and bureaucrats who carry out their professional tasks mechanically".

It is difficult to define the quality of life (QoL) of each patient and his family, since a myriad of subjective elements intertwines. The debate is often presented as the dilemma between life as a sacred value and QoL. Human life has a sacred value, but care must be taken so that it is quality life: not all the treatments that prolong biological life are humanely beneficial for the patient, and moreover, people are under no obligation to accept disproportionate means in order to prolong life.

Nevertheless, what is the dignity of human life?

According to a dictionary, the dignity of people, or their lives or activities means that they are valuable, and worthy of respect. Of course, there might exist consensus about what a dignified life or dignified death is like. Many people may agree as to what kind of life is respectable. However, what is the most important is that an individual perceive his/her life important and valuable. For example, suppose that many people regard living with a certain disability as undignified; but it is certainly possible that a person with the disability perceives his or her life important and worthwhile to live. In contrast, some may not tolerate a certain QoL, and feel undignified even if many others think it acceptable. Therefore, it is argued that the dignity of life has to be determined by an individual who lives the life as far as such determination does no harm to others.

The term ***"quality of life"*** references the general well-being of individuals and societies.

Within the field of healthcares, QoL is often regarded in terms of how it is negatively affected, on an individual level, a debilitating weakness that is not life-threatening, life-threatening illness that is not terminal, terminal illness, the predictable, natural decline in the health of an elder, an unforeseen mental/physical decline of a loved one, chronic, end-stage disease processes.

When patients cannot recover to decisional capacity and self-determination, we can protect their interests - and also their QoL- by empowering their surrogates to make decisions based on reasonable and attainable standards of evidence. This empowerment is justified because the DOC may be a paradigmatic example of a condition in which many persons would not want to be maintained, and because DOC patients have the same principled right to die that is held by both unconscious and conscious, legally competent patients. Obstructing the effective exercise of that right by DOC patients and their surrogates is a bias, a default position adopted in the void, which fails to consider whether continued life is desirable or in their interests.

Conclusion

With the notable advances of medicine, doctors are increasingly faced with ethical issues. The VS is just one of the many clinical conditions that obligate health professionals to reflect on ethical matters. A law that could limit ambiguity about the representation of the patient's will when in a VS in the matter of medical care is still missing. The Italian government is 'working on it', but there is concern about this is going to limit people's personal autonomy, and constrain living wills to some extent. Advanced care directives are not yet recognised under Italian law but probably represent the best way to safeguard the principle of autonomy. The trust between patient and doctor - the so called therapeutic alliance - is well put only when it enforces autonomy. We should not aim, paternalistically, at relieving the burden of sorrow, but rather inform and enable patients to fully understand, evaluate and make their own choices. Should a formal living will be unavailable, an opportunity to reconstruct the will of the incapacitated patient about ANH and life-supporting care withdrawal with the help of family members and friends should not be missed. There is an increasing need to calibrate accurately our interventions, trying to discriminate what is 'proportionate' and what is not, with regard to each one's own life experience and values rather than to a merely biological concept of life [30]. To do this, technical expertise is not sufficient; other factors, more personal and difficult to standardise, concerning the meaning of life itself and its quality, come into play. Physicians have a great responsibility to use their knowledge and skill in the primary interest of the patient. Magistrates are asked to make a great effort to interpret the rules, whereas politicians should take cognisance of the perverse effects the invasion of individual life sphere may cause, when biased moral conceptions are imposed on an entire society. It is urgent and necessary that the government acts, on one hand, by giving legal value to advance directives, and, on the other hand, by issuing a law that allows physicians to withdraw life-sustaining treatments without the risk of incurring criminal punishment. Finally, those who have the duty of informing and educating the public opinion should help people to cope with the finiteness of human condition, the intrinsic limits of medicine, and the responsibility to give in due time indications on values inspiring the choices about end-of-life.

References

[1] Bernat J. L. Chronic disorders of consciousness. *Lancet.* 2006; 367 (9517):1181-92.

[2] Strens L. H, Mazibrada G, Duncan J. S, et al. Misdiagnosing the vegetative state after severe brain injury: the influence of medication. *Brain Inj.* 2004; 18 (2):213–8.

[3] Rubin E. B, Bernat J. L. Ethical aspects of disordered states of consciousness. *Neurol. Clin.* 2011; 29 (4):1055-71.

[4] Owen A. M, Coleman M. R, Boly M, et al. Detecting awareness in the vegetative state. *Science* 2006;313(5792):1402.

[5] Fins J. J. Neuroethics and neuroimaging: moving toward transparency. *Am. J. Bioeth.* 2008; 8(9):46–52.

[6] Fins J. J, Illes J, Bernat J. L, et al. Neuroimaging and disorders of consciousness: envisioning an ethical research agenda. *Am. J. Bioeth.* 2008; 8 (9):3–12.

[7] Rubin E. B, Bernat J. L. Consent issues in neurology. *Neurol. Clin.* 2010; 28 (2):459-73.

[8] Tarquini D, Congedo M, Formaglio F, Gasparini M, Marcello N, Porteri C, Pucci E, Zullo S, Defanti C. A. Persistent vegetative state: an ethical reappraisal. *Neurol. Sci.* 2012; 33 (3):695-700.

[9] The Quality Standards Subcommittee of the American Academy of Neurology. Practice parameters: assessment and management of patients in the persistent vegetative state (summary statement). *Neurology* 1995; 45 (5):1015–8.

[10] The Multi-Society Task Force on PVS. Medical aspects of the persistent vegetative state (2). *N. Engl. J. Med.* 1994; 330 (22):1572–9.

[11] Bernat J. L. Ethical aspects of determining and communicating prognosis in critical care. *Neurocrit. Care* 2004; 1(1):107–17.

[12] Larriviere D, Bonnie R. J. Terminating artificial nutrition and hydration in persistent vegetative state patients: current and proposed state laws. *Neurology* 2006; 66(11):1624–8.

[13] Bacon D, Williams M. A, Gordon J. Position statement on laws and regulations concerning life-sustaining treatment, including artificial nutrition and hydration, for patients lacking decision-making capacity. *Neurology* 2007;68(14):1097–100.

[14] Bloche M. G. Managing conflict at the end of life. *N. Engl. J. Med.* 2005; 352:2371-73.

[15] Merrell D. A. Erring on the side of life: the case of Terri Schiavo. *J. Med. Ethics*. 2009 May;35(5):323-5.
[16] Quill T. E. Terri Schiavo-a tragedy compounded. *N. Engl. J. Med.* 2005; 352: 1630-33.
[17] Kahane G, Savulescu J. Brain damage and the moral significance of consciousness. *J. Med. Philos.* 2009; 34:6–26.
[18] Schneiderman L. J, Jecker N. S, Jonsen A. R. Medical futility: its meaning and ethical implications. *Ann. Intern. Med.* 1990;112(12): 949–54.
[19] Jecker N. S, Schneiderman L. J. Futility and rationing. *Am. J. Med.* 1992; 92 (2): 189–96.
[20] L Syd M Johnson. The right to die in the minimally conscious state. *J. Med. Ethics* 2011; 37:175-178.
[21] Striano P, Bifulco F, Servillo G The saga of Eluana Englaro: another tragedy feeding the media. *Intensive Care Med.*2009; 35:1129-31.
[22] Hurst S. A, Mauron A. Assisted suicide and euthanasia in Switzerland: allowing a role for non-physicians. *BMJ*. 2003; 326:271–273.
[23] Lindblad A, Juth N, Fu¨ rst C. J, Lyno¨e N. When enough is enough; terminating life-sustaining treatment at the patient's request: a survey of attitudes among Swedish physicians and the general public. *J. Med. Ethics*. 2010; 36:284–289.
[24] Mu¨ller-Busch H. C, Oduncu FS, Woskanjan S, Klaschik E (2004) Attitudes on euthanasia, physicianassisted suicide and terminal sedation–a survey of the members of the German Association for Palliative Medicine. *Med. Health Care Philos.* 7:333–339.
[25] Lemaire A. Law for end of life care in French. *Intensive Care Med.* 2004; 30:2120.
[26] Emanuel E. J. Euthanasia and physician-assisted suicide: a review of the empirical data from the United States. *Arch. Intern. Med.* 2002; 162:142–152.
[27] Bu¨low H. H, Sprung C. L, Reinhart K, Prayag S, Du B, Armaganidis A, Abroug F, Levy M. M. The world's major religions' points of view on end-of-life decisions in the intensive care unit. *Intensive Care Med.* 2009; 34:423-430.
[28] Solarino B, Bruno F, Frati G, Dell'erba A, Frati P. A national survey of Italian physicians' attitudes towards end-of-life decisions following the death of Eluana Englaro. *Intensive Care Med.* 2011;37(3):542-9.
[29] Barry R, The Papal Allocution on Caring for Persons in a "Vegetative State" Issues in Law and Medicine 20. 2 (Fall 2004): 155-64.

[30] Luchetti M. Eluana Englaro, chronicle of a death foretold: ethical considerations on the recent right-to-die case in Italy. *J. Med. Ethics*. 2010; 36 (6):333-5.

Index

B

C

D

E

F

I

J

K

L

M

N

O

P

Q

R

S

T